# COMMUNITY ENERGY

D1331463

BOROUGH OF POOLE

551317052 T

# Acknowledgements

I'd like to thank the following people for their help with writing this book:

Chris Balance, Vijay Bhopal, Matthew Black, Tommy Black, Richard Body, Alan Brown, Johannes Butscher, Ian Cooke, Katherine Cowtan, Clare Freeman, Werner Frohwitter, Pauline Gallacher, Chris Gathercole, Kate Gilmartin, Michael Groves, Nick Gubbins, David Gunn, Angus Hardie, Evelyn Hay, Anne Henderson, David Howell, Kayt Howell, Caroline Julian, Sue Kearns, Ian Kerr, Eleanor Logan, Julian Luttrell, Andrew Lyle, Kelly McIntyre, Deborah Macken, Steve Macken, Chris Morris, Laura Nicholson, Rachel Nunn, Sarah Peters, Stephen Phillips, Rowena Quantrill, Stuart Reid, Mark Ruskell, Emily St Denny, Anne Schiffer, Rose Seagrief, Peter Skabara, Maf Smith, Carolyn Somerville, Jan Szechi, Alis Taylor, Andrew Thompson, Ross Weddle, Anne Winther.

Particular thanks are due to Rachel Coxcoon from the Centre for Sustainable Energy in Bristol, and, of course, Ada Coghen, Niall Mansfield and all the team at Green Books for their patience in coaxing me through this process.

# COMMUNITY ENERGY

## A GUIDE TO COMMUNITY-BASED RENEWABLE ENERGY PROJECTS

Gordon Cowtan

green books

Published by
**Green Books**
An imprint of UIT Cambridge Ltd
www.greenbooks.co.uk
PO Box 145, Cambridge CB4 1GQ, England
+44 (0) 1223 302 041

Copyright © 2017 UIT Cambridge Ltd
All rights reserved.
Subject to statutory exception and to the provisions of relevant
collective licensing agreements, no part of this book may be
reproduced in any manner without the prior written permission of
the publisher.

First published in 2017, in England.

The Author has asserted his moral rights under the Copyright,
Designs and Patents Act 1988.
Illustrations © Gordon Cowtan
Cover design by Stephen Prior
Cover photographs by Westmill Solar Co-operative, Peter Skabara
and Katherine Cowtan
All photographs not otherwise credited © Gordon Cowtan
The publishers have endeavoured to identify all copyright holders,
but will be glad to correct in future editions any omissions
brought to their notice.

ISBN: 978 085 784 249 7 (paperback)
ISBN: 978 085 784 250 3 (ePub)
ISBN: 978 085 784 251 0 (pdf)
Also available on Kindle

Disclaimer: the advice herein is believed to be correct at
the time of printing, but the author(s) and publisher accept
no liability for errors or for actions inspired by this book.

10 9 8 7 6 5 4 3 2 1

# CONTENTS

# PREFACE

This book is split into four parts. The first part provides an introduction to community energy, giving information on what community energy is and why it matters. The second part covers the practical aspects of the different technologies like solar power and wind power, and looks at what's necessary to get a successful project off the ground. Part 3 is all about the gritty detail of legal structures and financial models for community energy projects. It also provides information and advice on other practicalities of projects like planning permission, sources of finance and sources of support. Finally, Part 4 looks at some of the reasons why so many community energy projects fail as well as considering what community energy might look like in the future.

Throughout the book I've tried to use as many case studies as possible. The different projects that come under the 'community energy' heading have such a broad range that they can be bewildering until they're grounded in what people have actually been doing.

# PART 1

# AN INTRODUCTION TO COMMUNITY ENERGY

Part 1 provides an introduction to community energy, firstly by describing what community energy is – the breadth of projects that communities undertake, what common themes there are, and where community energy as a movement stands now. It moves on in Chapter 2 to look at why community energy matters – what its role is in addressing climate change and improving sustainability, how it can earn money for communities and briefly looking at how far it can go. Chapter 3 sets this in the context of the history of the movement, tracing it back to the 1970s when how we use energy first began to be questioned, through the growing awareness of climate change and then looking at how government policies since the 1990s have affected the growth of community energy. Finally, Chapter 4 takes a detailed look at government policies – how the UK government finally recognised community energy, what policies there have been to support the movement and also what policies the devolved governments in Scotland, Wales and Northern Ireland have put in place.

Chapter 1

# WHAT IS COMMUNITY ENERGY?

Community energy is often seen as simply communities putting up wind turbines or building hydroelectric schemes, and while this is an important part of what it's about, there's a lot more going on in the range of activities that community energy groups are involved with and the effect and importance these activities can have.

## A few examples

The best way to get a sense of what community energy is all about is to look at a few examples.

### Isle of Gigha

One of the best known examples of a community buyout in Scotland, the Isle of Gigha - the southernmost island of the Hebrides - is also a community energy pioneer. By owning and operating its own wind farm, the Isle of Gigha was one of the first communities in the UK to embrace community energy. Known as The Dancing Ladies (Creideas, Dòchas, and Carthannas, Gaelic for Faith, Hope and Charity), the three second-hand Vestas V27 turbines were erected in 2003 and have operated since then, bringing in up to £100,000 per year to the local community, which at the time the turbines

were built had just bought the island from the previous owner.

Like most island communities off the west coast of Scotland, Gigha had been in decline for many years when in 2001 the owner put the island up for sale. Despite an asking price of nearly £4 million, the islanders organised a community buyout, forming the Isle of Gigha Heritage Trust as the legal vehicle to make this happen.

However, the buyout was just the beginning of things for Gigha. There was a strong feeling that the island had been in the doldrums and that life could be better. Gigha's location means that, like most of the west coast of Scotland, it receives its fair share of the weather. Using wind turbines was one of the ways chosen to improve life, harnessing some of this weather to generate an income stream for the island.[1]

Speaking about the wind turbines, Willie McSporran, the first chairman of the Isle of Gigha Heritage Trust, said, "Until recently the Isle of Gigha was in decline with a dwindling population and economy. When Gigha's community bought the island, we realised we needed to develop in a sustainable way and that is what our three 'Dancing Ladies' are helping us to do. Our three wind turbines generate electricity and an income for the community and they help to protect one of the island's greatest assets – the environment. Gigha's community is small but we are making a difference. If every community acts on climate change then we can solve it."

**One of the turbines on the Isle of Gigha.**
(photo: Dr Eleanor Logan)

More recently, a fourth turbine has been added and the island is also now part of an electricity supply experiment looking at how large-scale battery technology combined with wind turbines can help maintain the electricity supply when the island is disconnected from the mainland grid, as occurs frequently to many of the islands off the west coast.

## Ashton Hayes

A few years after Gigha took its groundbreaking steps into community energy, the village of Ashton Hayes in Cheshire started its first involvement in community energy, though in a very different way.

As the Ashton Hayes website states, "The Ashton Hayes Going Carbon Neutral Project is a community-led initiative that is aimed at making our village the first carbon neutral community in England."[2] The Ashton Hayes project launched in January 2006, when around 400 people came to a meeting in the village primary school. This meeting followed the parish council adopting a proposal by one of the residents of the village, Garry Charnock, that the village should try to become carbon neutral. Over the years since the project launched, the community estimate they have cut their carbon dioxide emissions by around 23%.

They've been supported along the way by a team led by Professor Roy Alexander from the University of Chester. One of the benefits of this involvement is that over a number of years, university students have carried out door-to-door surveys collecting information about aspects of people's lives that will affect their carbon footprint. This information has been collated and used to calculate the carbon footprint for the whole village. On top of this, residents have also been given information and guidance on what steps they can take to cut their carbon emissions further.

More recently, the village has set up a community interest company to build up community-owned renewable energy generation that will create profits that can be ploughed back into the local community.

## Fintry

The village of Fintry, just north of Glasgow in Scotland, took yet another approach to community energy. In 2003, a commercial wind farm developer presented its plans for a wind farm consisting of 14 2.5 megawatt (MW) turbines on the nearby Fintry Hills. Members of the community believed that this was an opportunity for the village, rather than a threat, and approached the developer with the idea of the village having a stake in the development, in the form of one of the turbines. Following much discussion and negotiation, the developer agreed to add an additional turbine for the village to the plans for the development.

The electricity from this turbine is sold to the grid along with the electricity from the rest of this development and the village receives the income this generates. An organisation to manage this income, Fintry Development Trust was set up specifically for the purpose.

**The village of Fintry in Stirlingshire.**
(photo: Gordon Cowtan)

The aims of the trust are to reduce the energy use and carbon footprint of the village as a whole, and use the income from the turbine to further these aims in the form of practical help for householders and other community groups in the village.

The wind farm has now been operating since early 2008. The trust has over 200 members from the village and has so far received around £450,000 from the operation of the wind turbine. The trust has also worked on a wide range of other activities, including:

- Installing insulation in 50% of the households in the village.

- Helping to install a woodchip biomass boiler in the local sports club to replace the existing oil boiler (see Chapter 9 for more information on using biomass as a heat-source).

- Employing energy advisors in the village to provide free, unbiased energy advice to householders.

- Helping with over 100 domestic renewable energy installations.

- Setting up a small biomass district heating scheme for 25 households.

- Facilitating the installation of external wall insulation on hard-to-treat homes.

- Running a large number of workshops for householders on subjects ranging from draughtproofing to beekeeping.

### Reepham

The three projects I've summarised above are all large and ambitious, involving large sums of money and multiple different initiatives, but community energy projects don't need to be like that. In 2004 in the market town of Reepham in Norfolk, the Reepham Green Team was established – "a network of individuals and community groups focused on developing projects to reduce the town's carbon footprint."

One of the Green Team's first actions was to undertake a comprehensive community carbon audit. A significant conclusion of this audit was that only 8% of the houses in the town were properly insulated. To tackle this the Reepham Insulation Project, known as RIP CO2, was set up. This project is described in more detail in Chapter 11.

As a rural area that is not connected to the mains gas supply and with poor public transport, many householders are reliant on oil to heat their properties and private cars for tansport. The initial audit in 2004 also indicated that the carbon emissions per person were 48% above the national average. To start to address this, as well as the RIP CO2 project, Reepham has also been the location of the first trials of biofuel in domestic boilers and have also set-up a local car club.

On top of this there has been work done on insulating the Town Hall, improving the energy efficiency of the primary school and setting up a wood fuel club in the local area.

Two aspects of the approach at Reepham are particularly interesting: firstly running projects that are replicable and, secondly involving other, already existing community groups in projects

### Repowering London

Based in South London, Repowering London is typical of a number of organisations that have started to appear around the country who aim to facilitate and act as catalysts for community energy projects. Repowering London works with community groups and local authorities to:

- Reduce $CO_2$ emissions by generating decentralised low-carbon energy.

- Tackle fuel poverty and educate residents about energy efficiency.

- Promote local leadership through cooperative community engagement.

- Provide opportunities for local and responsible financial investment.

- Create training and employment opportunities for local people.

Based within the communities it works with, it provides knowledge and expertise for projects that otherwise might struggle to get off the ground and also helps with raising the finance required to make projects work. So far it's been instrumental in three solar cooperatives in the Brixton area, with a total of over 130kW of photovoltaic panels (see page 14) being installed on roofs, and are currently actively working on four more projects.

As well as facilitating energy generation projects, Repowering London also provides work experience programmes for estate residents with each community energy project. These programmes range from a 30-week programme for ages 14–24, an adult 8-week programme and a 2-week programme for people who would like to learn about the installation of solar panels.

### Linlithgow

The Scottish town of Linlithgow has a population of around 14,000 and lies about 20 miles west of Edinburgh. The town has a very active Transition group* and from 2010 to 2012, with the support of the Scottish government's Climate Challenge Fund (see Chapter 4), this group ran two schemes promoting Solar Thermal and Solar PV.

The aim of the schemes was to facilitate the installation of solar panels by householders. This was done by the

**Solar installation in Linlithgow, West Lothian, Scotland.**
(photo: James Hiddinga, Transition Linlithgow)

---

\* A Transition group is a grassroots community project that seeks to build resilience in response to peak oil, climate destruction, and economic instability.[3]

group identifying the best value equipment and installers, so that each householder didn't have to go through this process for themselves, and an effort was made to source equipment and installers that were local.

Using the Climate Challenge Fund grant, the group were able to hire staff, who could then guide householders through the process and perform pre-installation checks so that the installers were confident that each house was suitable for the intended installation.

Over the two years that the schemes ran, around 200 solar thermal and 200 solar PV installations took place in the town.

## Common themes

Community energy therefore covers a wide variety of activities. This diversity of projects can make it difficult to pin down exactly what's meant by the term, though there are some common themes:

### Energy

It's pretty obvious that energy is always part of community energy projects, but even here there's a remarkably wide variety of different ways that issues to do with energy can be involved. The Gigha, Fintry, Repowering London and Linlithgow examples all involve the generation of electricity using wind turbines or solar panels. This type of project, a community owning the means of generating energy, is probably what people often first think of as community energy.

With their wind turbines, Gigha and Fintry are generating electricity on a fairly large scale and using the income this produces within the local community for other projects. Gigha is using the income for varied projects within the island while Fintry uses its income directly for other energy-related projects.

Using solar panels, Repowering London and Linlithgow are generating electricity (and heat) as a means of directly addressing the amount of energy households are using and, as a consequence, how much they're spending on fuel bills. In the case of Repowering London, they are also engaging local investors in the schemes, who then receive an income on their investment.

Ashton Hayes and Reepham are also addressing individual energy use, but rather than doing this via generation, they aim to reduce the amount of energy that's being used by improving energy efficiency and, particularly in the case of Ashton Hayes, encouraging people to change their behaviours so that they use less energy.

So, even though energy is always a vital element of community energy projects, there are a wide variety of forms that this can take and how people benefit from it. Looking at the examples above, there are three different ways energy can be involved in community energy:

- Energy generation – how heat or electricity is generated.

- Energy reduction – reducing the energy used by individuals or communities by improving insulation for example.

- Behaviour change – changing the way people think about and use energy.

## Community

As with "energy", it's a statement of the obvious that communities ought to be involved in community energy projects but, just like energy, there's slightly more to the community element than might appear to be the case at first glance.

The obvious part of community is that the projects are based in a geographic community of some sort: in the examples above, there are villages, a market town, an island and an area of London as the communities within which the projects take place.

There are some less obvious characteristics as well though. Firstly, the organisations that run the projects are answerable to the communities that they're part of through democratic structures. This usually means that they're run by some type of membership organisation – a co-operative or a company limited by guarantee – that people in the local community can join. Of the examples above, the projects in Linlithgow, Fintry, and Gigha are all run by organisations that are formally democratic in how they're constituted, while the co-operatives that Repowering London helps to set up are also by their nature democratic. There's more information about organisation types and legal structures in Chapter 13.

Secondly, these organisations are inclusive. This means that anyone in the local community who agrees with the aims and objectives of the organisation should be able to join. Community energy projects work best when they are part of the community that they're in. The dynamics of how communities work is a subject on its own, but any hint of exclusivity within a community can cause problems. Part of the democratic nature of the organisations mentioned above is that the mechanisms for enabling this democracy are inclusive.

Thirdly, the organisations encourage active participation in their projects. This means that people in the local community can get directly involved with projects. This level of participation is going to vary a lot depending on both the individuals and the organisation. For some it will be a matter of attending annual general meetings, while for others it will be spending significant amounts of time making projects happen. In the case of Repowering London, local people can attend training courses on all aspects of running community energy projects, including learning practical installation skills. Reepham directly engage with other community organisation and involve their membership with what they're trying to achieve. One of the characteristics of community energy is that projects are being done *by* the community not being done *to* the community. This "bottom-up" approach is vitally important to the success of any project.

## Philanthropic

Community energy projects are inherently philanthropic. All the projects that I described above are by their nature looking to share their benefits with the wider community and not just a select few. For example, both Fintry and Gigha use income generated from their wind turbines to support other activities in their local communities, while Linlithgow and Ashton Hayes look to reduce the amount of money people are spending on

their fuel bills. As well as directly benefiting householders and investors, Repowering London seeks to provide training and new employable skills for local people as well.

## Sustainability

Somewhere either in the background or the foreground of nearly all community energy projects is a desire to improve the sustainability of local communities and help with regeneration projects. This is probably most obvious with the Isle of Gigha, where an island community that had been failing for a number of years was given a new lease of life when the community bought the island from the previous owner and started to look for ways to improve life on the island to make it a better place to live with a desirable future.

However, the engagement with sustainability and regeneration is generally broader than just focusing on the immediate local community. As the Transition Linlithgow website states,[4]

> Transition Linlithgow (TL) is a community-led charity, which aims to make Linlithgow a more caring, sharing and resilient place to live with a focus on finding a beneficial pathway towards a low-carbon future.

Nearly all community energy projects have a desire to tackle climate change by modifying our use of energy at a local level. The name Ashton Hayes Going Climate Neutral embodies these aspirations.

Many community energy projects are initiated by Transition organisations. The Transition movement was born in Devon in 2006, growing out of work that Rob Hopkins had been doing for the previous 15 years[3]. It has since spread widely throughout the UK and many other parts of the world. It's based on the need to create communities that will be resilient and able to cope with the shocks that will occur as a result of climate change and the potential of dwindling and increasingly expensive fossil fuel resources. There are now a huge number of Transition initiatives, individual groups that have set up and are undertaking projects to make their local areas more resilient. The Transition Network[5] currently lists almost 500 official initiatives.

In this chapter, only the Linlithgow project is run by a group that identifies as a Transition organisation, but many of the community energy projects around the UK have been started by Transition organisations or have key members who are also members of Transition organisations.

## Opportunism and innovation

Many community energy projects in the UK often incorporate a significant amount of opportunism and innovation. This isn't opportunism in the negative sense of taking selfish advantage of circumstances, but simply a community group spotting an opportunity to make something happen that other organisations are not in a position to be able to exploit.

For example, the Fintry project spotted that there was an opportunity to do a deal with a commercial organisation that would be for the benefit of the local community. Similarly, Linlithgow spotted that with the introduction

of the Feed-in Tariff* for solar panels, a neutral, unbiased organisation would be able to engage with householders and achieve a high uptake of the technology. Gigha spotted that there were innovative ways of financing a community-owned wind farm and that the wind farm could help support the island.

This opportunism and innovation is at least in part a result of community energy not being part of mainstream energy strategy for governments in the UK, so communities have had to find spaces to do their own thing. The role of government in community energy is discussed in more detail in Chapter 4 but it would be fair to say that the movement grew *despite* government involvement or help rather than *because of* it (although it's also fair to say that successive Scottish governments have been more supportive).

It's possible that as the sector matures and becomes established, the innovative characteristic will become less noticeable.

## Multiple projects

One of the great things about community energy is that it can be the gift that keeps giving. All the organisations I've used as examples in this chapter have been involved in multiple initiatives and projects after the first one. The insulation project described is only one of over 40 projects that the Reepham Green Team has been involved with over the years, including a local car club and biofuel trials; since Gigha's initial three wind

---

\* The Feed-in Tariff or FIT is a government scheme designed to encourage the uptake of smaller-scale renewable electricity generation technologies. There's more information on how the FIT works in Chapter 12.

turbines were built, it has built a fourth and is part of a ground-breaking energy demand experiment; setting-up multiple projects is the reason for Repowering London's existence; and as befits a Transition Town, Linlithgow have been involved in a wide range of projects including allotments, orchards, timber recycling and a car club.

This willingness to continue embarking on new projects and taking on multiple initiatives is not just limited to the examples I've used in this chapter but is typical of community energy projects. Very often, the projects are interlinked as well, like Fintry, where the income from the wind turbine helps to pay for other energy projects in the village.

### A team

Finally, another, apparently vital characteristic of many community energy projects is that at their heart is a small team of highly committed people. This team can be as small as one person but more typically there are up to five or six people. The term "community energy" can suggest that these projects come about as a result of a groundswell within a community – armed with pitch-forks and fire, the people rise up and demand an energy project. In fact the truth is significantly more prosaic. While there's very often strong support from within a community, there is usually a "hard core" of people who have the vision and determination to make it happen.

This is true of Fintry, where four people gained the approval of the community council, negotiated with the wind farm developer and set up the development trust. From talking to other people from other projects, this pattern is repeated and is one of the most distinctive threads that runs through nearly all the projects – there are clearly a small number of prime movers who made Ashton Hayes Going Carbon Neutral happen and similarly with the Reepham Green Team and Repowering London.

This book is not intended to provide a recipe for how to run your own community energy project. That recipe doesn't exist, but the one ingredient that I don't think any project can do without is this team of people at the centre of it who are also part of their local community. Without this, any project is going to struggle to succeed; with it you have a chance of success.

## A few statistics

Trying to get a clear idea of how many community energy projects and organisations there are is very difficult. According to the UK government[6] at least 5,000 community groups have considered, commenced or completed energy projects in the UK since 2008. It's not possible to quantify how many of the 5,000 are currently active but even allowing for that, 5,000 is a huge number given that 10 years previously, there wouldn't have been more than a handful. Even more impressive is the evidence suggesting that a substantial number of these groups entered the sector in the last three years.

There are now groups all across the UK, though the numbers in Scotland and the south-west of England are significantly higher per person than in other parts of the country, and there is more activity in rural areas per person than in urban areas.

## Community energy generation

There's also no UK-wide government register of community energy generation projects – there's no requirement for community energy generation projects to notify anybody of their existence. The following are figures that have been put together painstakingly by Scene in Edinburgh[7] and were correct up to August 2014:

| Type | No of projects | Total MW | MW community owned |
|------|------|------|------|
| On-shore wind | 50 | 315.78 | 46.605 |
| Hydro | 18 | 1.986 | 1.986 |
| Solar | 36 | 7.7459 | 7.749 |
| Biomass | 7 | 1.242 | 1.242 |
| **Total** | **111** | **326.76** | **57.58** |

Table 1: Total number of generation projects with direct community involvement.

| Type | England | Scotland | Wales | N. Ireland |
|------|------|------|------|------|
| On-shore wind | 9 | 38 | 2 | 1 |
| Hydro | 10 | 6 | 2 | 0 |
| Solar | 26 | 9 | 1 | 0 |
| Biomass | 7 | 7 | 0 | 0 |
| **Total** | **45** | **60** | **5** | **1** |

Table 2: Generation projects by country.

Table 2 above suggests that the number of generation projects in Scotland per head of population is far more than would be expected. This is primarily for two interrelated reasons: firstly Scotland generally has the best resources for renewable energy generation – it is, on average, windier and wetter than most of the rest of the UK (though not sunnier!). This is reflected by there being more wind and hydro projects in Scotland per head of population than in other parts of the UK. The second reason is that the Scottish government was relatively quick to recognise that community energy might have a role to play and was the first to introduce support for what at the time was more or less a non-existent sector.

## The how and the what

As the preceding sections suggests, it's fair to say that *how* community energy projects are delivered is as significant as *what* community energy delivers. The issues of community energy initiatives being democratic, inclusive, philanthropic, and bottom-up are significant aspects of what community energy is and are important factors in why it has the potential to be a powerful and important sector in the future.

Chapter 2

# WHY COMMUNITY ENERGY MATTERS

Community energy matters because it has the capacity to help address a number of the biggest issues we face as a society - issues like energy use, fuel poverty, sustainability, and climate change - and address these in a way that benefits the local community and the people who live there. In a report, Friends of the Earth Scotland summarised the benefits of community energy:[1]

> As community energy grows and develops as a concept, its benefits are better understood. In addition to helping achieve emissions reductions, it enables communities to harness local natural resources to build social capital, create local and regional employment opportunities, create revenue to address community development needs and combat fuel poverty.
>
> Community ownership and participation in projects can also help generate support and acceptance of renewables more broadly. Furthermore, involvement in community projects helps stimulate citizen interest in other areas of energy such as energy conservation and demand side management.

## The environment and climate change

For many people involved in community energy, minimising environmental damage is one of the reasons they're interested and involved. The environmental impact that concerns most people is climate change.[*] Burning fossil fuels is a major contributory factor to climate change and other means of satisfying our need for energy need to be found.

The following table shows the $CO_2e$[†] emissions as a result of producing a kilowatt-hour (kWh) of electricity from various different sources. These figures are taken

---

[*]  Interestingly, scientists have been aware of the potential of climate change for well over 100 years. The Irish physicist John Tyndall identified the "greenhouse effect" in 1861 (the Tyndall Centre was much later named in his honour) and in 1896 the Swedish chemist Svante Arrhenius concluded that burning coal enhances the natural greenhouse effect, while as far back as 1938 the British engineer Guy Callendar was the first to demonstrate that there was a correlation between rising global temperatures and rising $CO_2$ levels in the atmosphere.

[†]  $CO_2e$ or Carbon Dioxide Equivalent is a measure used to compare the emissions from different greenhouse gases based on their global warming potential. In the context of this book, it is used to measure the total global warming potential of the mixture of gases emitted from different processes, allowing those processes to be compared.

from the IPCC climate report[2] and include the full life cycle of building the plant as well as the actual generation of electricity:

| Generation technology | Total life cycle emissions (grams of $CO_2$e per kWh of electricity) |
|---|---|
| Coal | 820 |
| Gas (combined cycle) | 490 |
| Hydropower | 24 |
| Rooftop solar PV | 41 |
| Wind | 11 |

Table 1: Life cycle emissions from different electricity generation technologies.

And here's a similar table for heat generation, with the figures taken from the the UK government's carbon emission conversion factors[3]

| Fuel | Emissions (grams of $CO_2$e per kWh of heat) |
|---|---|
| Natural gas | 185 |
| LPG | 215 |
| Heating oil | 247 |
| Coal | 340 |
| Logs | 12 |
| Woodchips | 12 |
| Wood pellets | 12 |

Table 2: Life cycle emissions from different heat generation technologies.

These are of course averages and for any particular installation the figures will vary depending on the precise circumstances, but they very clearly demonstrate that from a greenhouse gas emissions perspective, the renewable fuels and technologies that are typically employed in community energy generation projects are significantly more sustainable than their fossil fuel counterparts.

This is why community energy generation projects are always using renewable technologies despite the UK government's Community Energy Strategy[4] raising the possibility that non-renewable technologies could be included within community energy.

## People like me

The other significant factor in the context of climate change is that community energy projects can normalise the introduction of energy reduction and renewable energy generation projects so that they are not something that "other people" do but something that "we" do.

Many people feel disengaged from climate change: it's large and scary but the effects are also, so far at least, distant and unspecific. It's rare to be able to identify and understand a direct cause and effect. Many of the steps required to address the issue also run contrary to much of our prevailing culture and the prevailing messages within our society, which has the effect of strengthening a sense of dissonance.

To make matters worse, the mechanisms used in the UK to shift capacity away from carbon-intensive generation towards low-carbon renewables have often alienated people who feel that their local environment and landscape is being changed by people and organisations that are distant and that they have no connection with (not even a retail connection because in the UK it has not been possible to sell electricity produced in an area to the people in that area).

It would be stretching things to claim that community energy can solve these issues but it has the potential to be a key ingredient in how we address them. Unlike commercial developments, community-owned renewable energy can strengthen local resilience far more than enhanced profit/dividend levels.

# Improving sustainability

"Sustainability" is one of those terms that has such a broad range of definitions that it can mean almost anything – or almost nothing.* The *Oxford English Dictionary* defines it as "able to be sustained or upheld at a particular level without causing damage to the environment or depletion of resources." Nearly all community energy projects involve either reductions in energy use or generating renewable energy so this definition works well.

In the context of community energy, the term works in another way as well, in the sense of improving the community itself by improving the lives of the people who live there and the cohesiveness of the community. Bringing money into a community organisation via a

---

\* The term "sustainability" really came to the fore after the Rio Summit in 1992, when one of the outcomes of the event was a "Declaration on Environment and Development" which incorporated the term "sustainable development" within many of its 27 principles.

renewable energy generation project, reducing fuel poverty, enabling people to invest money in their local community, or providing a focus for local people to work together are all means of improving life in a local area.

## Case study: Repowering London

As described in the previous chapter, Repowering London promotes and facilitates the development and local ownership of renewable energy projects across south London and is a good example of how community energy can work in the round to improve the sustainability of a local area.

The projects they run, fitting solar PV panels to the roofs of housing in south London provide a number of benefits:

- People who live in the housing benefit from reduced electricity bills as a result of the electricity generated by the panels.

- Local people can learn new skills by being trained in the installation of the panels.

- Other local people can invest money in the cooperatives knowing that their investments are benefiting their local community.

- The electricity produced is renewable and will be reducing the amount of electricity generated from non-renewable sources.

- They engage people who might otherwise not have any sense of engagement with how we tackle climate change.

## Sustaining communities

One of the great strengths of community energy is that it satisfies both sides of sustainability. As discussed above, it can reduce costs and generate income for community bodies, helping those bodies to continue and thrive, and it can do the same for individual households, helping to lift people out of fuel poverty among other things. It can also generate local jobs, not just for the projects themselves, but in helping to create and support supply chains of, for example, logs or woodchips.

It's worth making a distinction between commercial developments and community developments. With many larger commercial renewable developments, the company behind the project will be outside the local area and very often these companies will be borrowing money to pay for the project from banks that can be anywhere in the world. The result of this is that the income generated by the project from government incentives like the Feed-in Tariff or Renewable Heat Incentive (see page 98), can end up providing minimal benefit locally, because the majority of the money goes to organisations that are outside the local area that is hosting the project. The opposite can be true for community developments where the majority of the income stays in the local area and can be recycled through other local projects so that it boosts the local economy multiple times.

## Case study: Neilston Development Trust

Neilston is a small town of around 5,500 residents in uplands to the south-west of Glasgow. Although there is evidence of the area being inhabited as far back as the 12th century, like many parts of central Scotland the town really grew in prosperity and population in the industrial revolution as factories and cotton mills grew around hydropower supplied by the River Levern.

However, again like much the of the rest of central Scotland, this prosperity diminished in the second half of the 20th century as the mills gradually closed and the town became more of a base for Glasgow commuters.

Although located in one of Scotland's most prosperous areas, Neilston has pockets of severe disadvantage. Scottish Index of Multiple Deprivation (SIMD) figures were calculated for every Neilston postcode using the 2009 figures.[5] These revealed among other things that 15% of the population live in an area that ranks in the worst 15% in Scotland on the full range of deprivation indicators.

Neilston Development Trust (NDT), which was founded in 2006, has become recognised as one of Scotland's leading community development trusts and in 2009 the trust published Neilston's visionary Renaissance Town Charter.[6] The town charter outlines a vision for the community, placing the town's goals in the context of local council plans and the Scottish government's Sustainable Economic Growth policy.

The resulting Neilston 2030 vision describes a pathway towards the creation of a sustainable, economically robust, well-planned and well-connected small town. The next step was to find a source of sustainable income to implement the community's vision.

The trust was aware of successful joint venture wind farm schemes between communities and commercial wind farm developers on the other side of Glasgow, in

Fintry, and was interested in following a similar route as a source of long-term funding for its plans. It was introduced to a commercial developer keen on joint ventures with communities. A potential site for a small wind farm near Neilston was identified in 2009 and planning consent for a four-turbine 10MW wind farm was granted in 2011.

The commercial developer had paid the up-front costs required to achieve planning consent but to secure their involvement in the project, the trust had to pay their share of the capital costs. The trust had the option of owning up to 49.9% of the project and raised sufficient funds for just over 28%.

The wind farm was opened in May 2013 and it's estimated it will generate the community over £10m over the lifetime of the project providing an income to sustain the trust on its path towards the 2030 vision.

## Earning or saving money

Together with environmental concerns, earning money or saving money is the most obvious reason why community energy projects get started. For almost all projects this will be in there somewhere as one of the aims. It can be:

- People in the community saving money through reduced heating bills as a result of an insulation project.

- People within a community investing in a local energy generation project like a hydro scheme or a wind turbine.

- The costs for maintaining a village hall being reduced as a result of a new heating system being installed or solar panels being installed on the roof.

- A community energy initiative erecting a wind turbine or taking a share in a commercial wind farm to generate an income, which is then ploughed back into other community projects.

Saving money, whether by individuals or community organisations, as a result of improving insulation is easy to understand – more heat is retained in the building so less is needed to maintain a comfortable ambient temperature. However earning money though renewable installations is less straightforward.

Firstly heating or electricity costs can be offset by a reduced cost for fuel or by generating your own electricity via solar photovoltaic panels. So a building that was previously heated by an oil boiler will save money when this is replaced with a woodchip boiler because the cost of generating heat from woodchip is currently cheaper than oil. (The other option for a poorly heated building of course is to spend the same amount but heat it better.) Similarly, a building that uses a lot of electricity, whether for heating or other reasons, can save money by solar photovoltaic panels being installed that offset the amount of electricity used.

Most renewable installations receive government support. In the case of installations that generate electricity, this is via the Feed-in Tariff (FIT). The FIT means that for every kilowatt-hour of electricity gener-ated, the organisation owning the generator will save money on the electricity used on-site, be paid for all the electricity generated, even if it's used on-site, and will also be paid an additional amount for any surplus electricity exported to the grid. The amount received varies significantly depending on the technology (wind, hydro, photovoltaic panels, etc.) and the size of the generating equipment. The FIT is described in more detail in Chapter 12.

Following the UK general election in 2015, the government made a number of announcements reducing the amount paid by the FIT considerably – particularly for solar panels and for wind. The full effects of this have yet to be felt but there will be definitely be some projects that were previously financially viable that no longer are.

It appears that the government has also withdrawn all support for future on-shore wind farms over 5MW in generating capacity. If this policy is maintained then it will mean an end to the larger schemes in which communities might have had a share of the profits.

For installations that generate heat rather than electricity, like wood-fuelled boilers or ground-source heat pumps, there's a similar scheme called the Renewable Heat Incentive (RHI). This pays the owner of the system an amount for every kilowatt hour of heat generated. Again, the amount received varies depending on the technology and the size, and to confuse matters further there are two versions of the RHI, one for individual households and one for community buildings and schemes, with different values and rules.

## Biomass case study

St Michael and All Angels' Church in Withington, Gloucestershire undertook a comprehensive review of how it used energy and then put in a number of measures to reduce overall energy consumption – both electrical and heat. These measures included the installation of low energy lightbulbs inside the church and reduced operating times for external floodlights, the installation of a solar photovoltaic array and the installation of a wood pellet biomass boiler replacing the existing oil boiler.

The old oil boiler had become unreliable and whether it was needed or not was being used every day just to maintain reliability. In fact the heat demand for the church was very low – only seven hours per week. Replacing the oil boiler with biomass was an obvious choice because there was an existing wet radiator system in the church and both types of boiler produce heat at a high temperature.

Because the weekly heat demand was so low, it was decided it wasn't necessary to have an automatic feed for the pellets, instead the pellets are loaded by hand from 10kg bags into a 500kg hopper which, when full provides enough for eight weeks of operation. This helped to keep the cost of the installation down.

The installation was a 38kW Froling P4 pellet boiler and the total cost of installation was just over £23,000. This was funded partly from a mixture of public funds and partly from private donations. It's estimated that the church is saving £1,100 per year in fuel costs and over 12 tonnes in $CO_2e$ carbon emissions as a result.

Because of the new boiler and the other changes made, it is believed to be the first carbon neutral church in the UK.

# Energy costs and fuel poverty

In the past 15 years, energy prices have increased dramatically. For example, the market price of crude oil had gone from around $30 per barrel to over $100 per barrel in 2014 and at times has been over $140 per barrel. More recently it dropped but has been creeping back up and at the time of writing is above $50 per barrel.

While nobody buys crude oil to heat their home, the costs of almost all forms of energy are currently tied to that of crude oil. This increasing price of energy has pushed more households into fuel poverty, and a community energy project that saves money for individuals can also be play a part in tackling fuel poverty.*

Reducing the number of households in fuel poverty is often quite rightly a target of community energy projects either by improving the insulation of households so that it takes less energy to heat the house or by improving the heating system. In both cases the aim is to reduce the cost of heating the house for the householders.

Community organisations are frequently best placed to address local fuel poverty issues: it will often be neighbours and friends who live in the local area who are

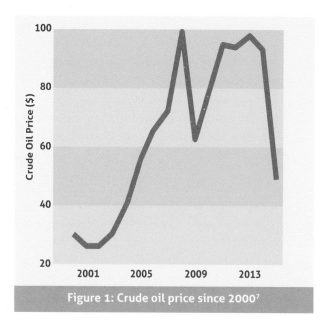

Figure 1: Crude oil price since 2000[7]

struggling with fuel bills, and community organisations can have a good understanding of the local area and how to engage successfully with local people. And, of course, many people simply like to improve life in their local neighbourhoods.

## Fuel poverty case study: Fintry Development Trust

As described in the previous chapter, although the primary aim of Fintry Development Trust is reducing energy use and environmental impact, soon after the organisation was founded a survey of energy use in the community found that over 40% of households were in fuel poverty.

Over the years since, the trust has undertaken a number of initiatives aimed at reducing this number. Among

---

* There are various definitions for fuel poverty, but the essence is that a household is in fuel poverty when the householders don't have enough money coming in to heat their homes and pay for other necessary items.

these, the trust has employed energy advisors who are there to provide unbiased advice to householders who want to take steps to reduce their energy use. These steps can range from the practical simplicity of fitting draughtproofing or insulation through to helping householders navigate through the labyrinth of government schemes that are available to support the installation of more efficient, low-carbon heating systems.

The trust has also run a number of specific projects over the years, including facilitating the installation of external wall insulation on hard-to-insulate homes and facilitating the installation of low-carbon forms of heating, like heat pumps and biomass boilers, for households that will benefit. Most recently, with assistance from the Scottish government the trust has installed a biomass, district-heating scheme for 25 park homes.

## How far can you go?

So, how far can community energy projects go? While there are many good ambitious projects in the UK, to see how far and how ambitious community energy projects can be, it's worth paying a visit to Feldheim, a small village not far from Berlin, in what was once part of East Germany.

Back in the days when Germany was divided, Feldheim was a communist collective farm. Now it's a living, breathing example of what can be achieved with community energy, why community energy can make a significant difference to the lives of ordinary people, and how community energy projects can play a part in tackling some of the biggest challenges we face.

The story starts back in 1995 when a local entrepreneur built the first wind turbine in Feldheim. This was followed by a 43-turbine wind farm built by a renewable energy company, Energiequelle GmbH, that supplied electricity to the grid and provided an income to local farmers in the form of land rents. After building the wind farm, Energiequelle then took over an old military site near the village and built a solar farm.

It was at around the same time as the solar farm was built that things got really interesting. The community wanted to use electricity that was produced locally and were shocked to find that they couldn't because the wind farm and solar farm sold their electricity to the grid, and the utility company that owned the local grid and

**Wind turbines and biogas plant at Feldheim in Germany.**
(photo: Werner Frohwitter, Energiequelle)

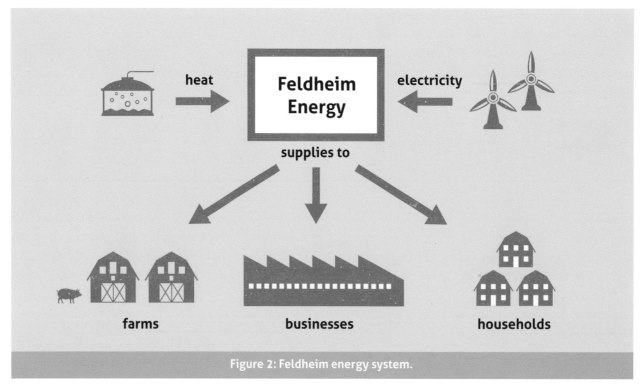

**Figure 2: Feldheim energy system.**

supplied electricity to the householders bought their electricity from another source far away. Initially, the people of Feldheim sought to buy the grid from the utility company but they didn't want to sell.

Feldheim decided the way around this was to build their own local electricity grid. In 2008 Feldheim Energie was formed – an independent utility company that's owned and managed by the people of Feldheim. The purpose of Feldheim Energy is to supply safe, locally produced, sustainable electricity and heat that's independent of the conventional utility-company grid. Feldheim Energy is now supplying electricity that costs around 30% less and

heat at around 10% less than people were paying previously.

They also built a biogas plant that takes pig manure from the local pig farms and corn from local arable farming and converts them into both electricity and heat (plus over 11,000 cubic metres of fertiliser as a by-product). As well as a local electricity grid supplying the locally produced electricity, Feldheim Energy also built a district heating system to supply the heat to local households and businesses.

To support the anaerobic digestion (AD) plant, Feldheim Energy installed a woodchip-fuelled boiler that cuts in when the demand for heat in the village is greater than the biogas plant can manage on its own. Similarly there are now plans to install large batteries to cope with the natural fluctuations in wind power production and electricity demand.

Feldheim is one of the best examples of how far community energy can go. What's particularly impressive about what's been done is that as well as controlling the systems that produce energy, the community also controls the systems that supply that energy to local households and businesses. By doing this, Feldheim demonstrates that community energy can play a part in addressing three of the issues that we all have to face with regard to energy use – energy costs, energy security and climate change.

## Energy costs

Because Feldheim is using renewable technologies to produce both heat and electricity, and because it is using only a fraction of what is produced, it can sell the remainder to the grid and is able to undercut the "market" cost of energy that the people and businesses in the village would have to pay otherwise.

What might be more important though, is that by taking the steps they have, the people of Feldheim are in control of how much this energy costs. By owning both the means of energy production and the means of energy distribution, and by using renewable technologies, how much the energy supplied in the village costs is not subject to the same external factors of oil prices and the activities of "the market". This means that there is more stability in the cost of energy over time, and for the

people living in the village a sense of having greater control over this aspect of their lives.

## Energy security

The issue of energy security is huge, covering everything from long-term investments required to supply energy inline with economic developments and environmental needs, to the ability in the short-term for an energy system to react promptly to sudden changes in the supply or demand of energy. It is fundamental to the economic health of a country – without energy an economy can't function – and many countries are looking for long-term solutions to increase energy security by reducing dependence on foreign oil and gas imports.

As well as being a concern for national economies, energy security is clearly also a local issue – households and businesses need to know they can rely on energy being available when they need it – and, in taking the steps it has, Feldheim has established local energy security. Energy is produced locally and using locally produced raw materials in the form of pig slurry and maize. On top of that, because of the local electrical and heat grids, it is this same energy that is distributed and supplied. They have even put in measures to deal with the peaks and troughs of energy supply and demand in the form of a biomass boiler for heat and the proposed storage batteries for electricity.

## Climate change

All of Feldheim's energy is generated using renewable technologies. The electricity is generated using wind turbines and biogas while the heat is generated using biogas and burning woodchip.

Not all communities are in a position to be able to do what Feldheim has done (for example, it's currently very difficult for communities in the UK to take control of their grids to sell locally produced electricity to households). The vast majority of community energy projects in the UK can only tackle parts of the jigsaw. However, Feldheim is a pointer to how community-owned energy companies could become part of a new landscape of energy partnerships delivering demand reduction, energy efficiency, energy storage and management, owned and operated by the local community and sensitive to that community's needs and the broader concerns of sustainability.

## Why community energy *should* matter

Although the number of community energy groups in the UK has been estimated as high as 4,000,[4] it's also true that the total capacity of community energy generation projects is not much more than 1% of all renewable generation in the UK, which itself accounts for only around 19% of total generation in the UK.[8]

In the light of these figures, it's worth asking not just why community energy matters now, but also why it *should* matter.

One of the leading lights of the community energy movement, Pete Capener,[9] described the effect of community energy projects as follows:

- I see 'People Like Me' involved, whether family, friends, neighbours, community members, work colleagues.

- I keep hearing about opportunities for involvement through many different local routes, it stops being unusual and becomes 'What Happens Around Here'.

- I trust the people delivering the projects, I see them around, they're local.

- I can see tangible benefits for my local area.

- I can see tangible benefits for me.

# Chapter 3
# HISTORY

Fittingly for a movement that at its best works organically from the ground up and pulls together a number of different strands, including energy generation, energy use and community development, it's not possible to write a simple linear history of community energy. Nor is it possible to identify a particular moment when community energy was invented or created by a specific government policy.

However, although government policy didn't set out to create the community energy movement, it's fair to say that the growth of the movement over the past 10 years is, at least in part, a consequence of government policy, particularly with regard to renewable energy generation.

## In the beginning...

As a recognisable movement, community energy is not yet much more than 10 years old and is still extremely fragile. However, like all movements that appear to emerge out of nowhere, it has roots that go much further back.*

---

\* It's probably stretching things a bit to claim it was an important element in the birth of community energy, but it's true that when the streetlights of Godalming in Surrey were converted from gas to electricity in 1881, this first 'public'

It's possible to trace the roots of community energy to the 1970s when a growing awareness of global issues, including population growth, energy resources and global warming, led to the birth and growth of radical environmental campaigning groups like Friends of the Earth and Greenpeace and the popularity of critiques of orthodoxies of the day like E.F. Schumacher's *Small Is Beautiful*.[1]

Concerns about energy resources in particular were amplified by the first oil crisis of 1973, which saw the price of crude oil quadruple and queues at petrol pumps. One of the responses to these global issues was to look to more local, decentralised models for social, political and economic structures; models in which communities would take greater responsibility for their own future.

## CAT and UCAT

Two organisations that are particularly relevant to community energy and embody many of the ideas and much of the ethos found in community energy projects were formed during this period. The first was founded in 1973 in an old quarry in Machynlleth, North Wales.

---

electricity supply in the UK was powered by what was also the first hydroelectric power station in Britain.

Originally known was the National Centre for Alternative Technology, it is now called the Centre for Alternative Technology, or simply CAT.

Founded as a community dedicated to eco-friendly principles and acting as a test-bed for new ideas and technologies and consisting of little more than a group of volunteers inspired by a vision but with no money, CAT has now grown to become one of Europe's leading eco-centres. One of the key decisions that was made early in its life, after a visit by the Duke of Edinburgh (of all people) in 1974, was to turn part of the site into a visitor centre to generate interest in the work it was doing. This visitor centre opened to the public in 1975.

CAT is now an educational charity that, among a huge range of activities, hosts school visits, runs short courses and postgraduate degrees, produces a quarterly magazine, *Clean Slate*, and provides environmental consultancy and renewable energy consultancy services. It has around 65,000 visitors each year and continues to work on new projects, schemes and ideas – looking for solutions to environmental challenges. The influence of CAT spreads much more widely than just those who visit or attend courses there – it has been a repository and advocate for many of the ideas, both technical and social, behind community energy projects.

The other relevant organisation was founded a few years later, in 1979. This sister to CAT had energy as a central theme but was also founded, in part, as a reaction to limitations inherent in CAT. Firstly CAT was built on a philosophy of rural self-sufficiency that would always have limited relevance to many people in the UK who lived in urban cities and towns rather than rural environments.

The second difference with this new organisation was a preference for "appropriate technology" rather than the "alternative technology" of CAT. The founders felt that any technical solution should be sensitive to the situation people are in rather than requiring a radical shift, which many people might have trouble engaging with. These two differences were embodied in the name of this new organisation – the Urban Centre for Appropriate Technology (UCAT).

**The UCAT Future City Home in Bristol, 1981.**
(photo: Centre for Sustainable Energy)

As well as having energy has a central theme, UCAT was also inspired by how social groups function successfully and by co-operative principles. From initial meetings of volunteers in 1979, the membership of UCAT grew quickly and it occupied its first premises, a house in Bristol, in 1981. The volunteers drew up plans for a "Future City Home" and rebuilt the house as the first ever low-energy rehabilitation of an old house in the

UK. In the 35 years since then, UCAT has grown, is still based in Bristol and is now known as the Centre for Sustainable Energy (CSE).

CSE manages innovative practical energy projects and also undertakes research and policy analysis on energy and related topics. Although it remains based in Bristol (and is one of the reasons why Bristol is a hub of community energy activity), like CAT, CSE's influence spreads far and wide. It has been important to the community energy movement, not just because of its focus on energy but because of a belief that to achieve its aims of cutting carbon emissions and reducing fuel poverty, it should "focus on enabling and supporting other people to take action within their own homes, communities and organisations".[2] CSE has therefore played a practical role in numerous community energy projects, particularly in the south-west of England, and in many other ways including advising and nudging successive governments.

Over the years, both CSE and CAT have provided information and inspiration for community energy projects with their focus on raising awareness of energy use, and localising and democratising energy production and distribution.

## Climate change

While CAT and CSE provided much of the inspiration and information that community energy projects grew from, it wasn't until the late 1990s and early 2000s that successful individual projects started to appear around the UK that viewed together could be seen to be starting to form a recognisable movement. These projects were as varied in their approach and size as they were in their locations. Among them were the Baywind Co-operative in Cumbria, Gigha off the west coast of Scotland and Fintry in central Scotland, Ashton Hayes in Cheshire and a number of community hydro projects including Settle and Torrs in Yorkshire and Derbyshire respectively.

There are a number of reasons why these projects should all have started to appear around the same time, but among the key identifiable ones include the level of media coverage and public interest in global environmental issues, specifically climate change and increased government support for renewable energy.

The global environmental issues of population growth and climate change that influenced the founders of CAT and CSE were all still (unfortunately) significant issues in the late 1990s. It was around this time that climate change in particular began receiving significant coverage in the media. As noted in the previous chapter, the phenomenon of climate change has been known for over 150 years, while the mechanisms that cause it have been understood for almost 120 years. However, it received very little media coverage until much more recently. There were occasional newspaper articles in the 1930s and 1950s about climate science and links to anthropogenic sources but, through the 1960s and 1970s coverage remained sparse.

This changed in the 1980s and in particular in the late 1980s when media coverage of the issue and public awareness increased significantly. The chart opposite shows how coverage increased over the years and how it correlates with releases of reports from the Intergovernmental Panel on Climate Change (IPCC) in 1990, 1995 and 2001, and also during the 1992 UN Convention on Climate Change and the 1997 Kyoto Protocol. The initial increase in coverage in 1988 coincided with NASA scientist James Hansen's testimony to the US Congress that "it is time to stop waffling so much and say that the evidence is pretty strong that the greenhouse effect is here".[3]

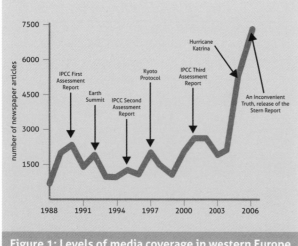

**Figure 1: Levels of media coverage in western Europe and North America.[4]**

This increase in media coverage over the years (which so far has peaked in 2007 with the release of Al Gore's film, *An Inconvenient Truth* and the publication of *The Stern Review* [5]) has led to an increase in public awareness of the issue and an increase in people wanting to take an active part in doing something about it.

## The NFFO years

This increase in media coverage and public awareness also led to governments wanting to address the issue and in particular increase the amount of energy generated from renewable sources.

Apart from the construction of significant numbers of hydroelectric power stations in the north of Scotland in the 1940s and 1950s,* the vast majority of the energy used in the UK for heating, lighting, transport and industrial processes since the industrial revolution had been generated from fossil fuels (plus a significant amount of nuclear power in the last 60 years). In the 1970s, provoked by many of the same factors that influenced the founders of CAT and CSE, the government invested a very small amount in renewable technology research and development.

However, in 1990 the UK government introduced the Non-Fossil Fuel Obligation (NFFO) and the Scottish Renewables Obligation. The NFFO was originally devised as a means of supporting nuclear power when the UK electricity industry was privatised in 1990 but it was extended to include renewable sources. It worked by increasing the amount that generators could get for electricity produced from renewable sources.

The first commercial wind farm built on the back of the NFFO was Delabole near the north Cornwall coast. The wind farm originally consisted of 10 turbines, each of which was rated at 400kW and became operational in December 1991. By September 2000, some 75 wind

---

* The story of Tom Johnston and the growth of hydro in the north of Scotland in the period after the Second World War is a fascinating one, with the approach the government took at the time contrasting with the approach of current governments to the growth of renewables. For more information, see Emma Wood's engaging account in *The Hydro Boys*. [6]

**Delabole wind farm in Cornwall**
(photo: Good Energy Ltd)

farms had been built in the UK with a total capacity of just over 170MW.

## Baywind

Some of the ingredients to enable practical community energy projects were now in place, including increased public concern about climate change and a government policy supporting the development of renewable energy generation.

The vast majority of these wind farms were built by landowners or commercial companies, with little or no involvement of local communities. One part of the country where this wasn't the case was in Cumbria, where an organisation called Baywind Energy Co-operative raised £1.2 million in 1996/97 to buy two turbines on the Harlock Hill wind farm and a further £670,000 in 1998/99 to buy one turbine at nearby Haverigg II.[7]

Baywind is a cooperative that was formed in 1996, emulating a model that had been successful in Scandinavia – Denmark in particular – where the energy

systems typically have a large number of small power stations rather than a small number of large power stations, as historically has been the case in the UK. Individuals become members of the co-op by investing money in it and the co-op is run on the democratic basis of "one member, one vote". (The co-op model is discussed in more detail in Chapter 13.) Much of the money raised by Baywind was from people who lived close to the developments or from the wider north-west of England region.

Over the years since its formation, members of Baywind have received between 3% and 10% on their investment each year with many of them receiving slightly more due to a tax relief called the Enterprise Investment Scheme (EIS).

Baywind's co-operative ownership of wind turbines was the first of its type in the UK, and following the success at the sites in Cumbria, a sister organisation called Energy4All was created in 2002 to enable the model to be copied in other communities around the UK. Over the years since then, Energy4All has helped to set up 20 co-ops that are owners or part-owners of renewable energy developments that now include solar developments and wood biomass as well as wind farms.[8]

Within the community energy movement there's often an argument about Baywind and Energy4All and whether or not they count as "real community energy", with some arguing that they provide little more than an investment opportunity. The counterarguments are that they make a real effort to raise as much investment as possible from people who live close to the developments; that the democratic co-operative model encourages all the investors to be active members rather than just investors; and that membership engages and involves people with where their energy comes from.

## The ROC years

With media coverage of climate change continuing to increase and with it public concern, the new government elected in 1997 reviewed the NFFO scheme. While it had been successful in reducing the cost of building renewable energy generation, the targets set for the amount of generating capacity had never been reached. The government replaced NFFO with Renewable Obligation Certificates (ROCs), which came into effect in 2001.

This new scheme had much higher ambitions for the amount of renewable electricity generation with a target of 10.4% by 2010 and an aspiration of 20% by 2020. It required electricity suppliers to purchase electricity that was generated from renewable sources and to demonstrate compliance through a tradable certificate scheme. The result was greater financial certainty and greater income for developers of renewable electricity schemes, with a consequent dramatic increase in the number of companies entering the market and, for wind farms in particular, the number of sites being considered for development.

The wind turbines on the Isle of Gigha, already described in Chapter 1, was one of the next major community energy developments. When the islanders had been contemplating a buyout one of the organisations that they contacted and that ultimately took a significant role in helping to make both the buyout and the wind turbines happen was Highlands and Islands Enterprise (HIE).*

HIE had already been aware of the potential for renewable energy to solve some of the funding challenges that community regeneration always seemed to face – volunteers often spending most of their time chasing grant funding rather than being able to progress their projects. It seemed that renewable energy could help solve this problem and, in echoes of the ethos of CAT and CSE, could solve it in a way that would have an environmental impact relating directly to the way that people behaved.

A Community Energy Unit was set up in HIE to facilitate the development of community renewable projects in the Highlands and Islands area. Initially this focused on relatively small projects – solar panels or small wind turbines for community buildings – that were straightforward and would give communities a tangible success that could lead to greater things if there was a sufficient need or desire within the community. The Community Energy Unit became Highlands and Islands Community Energy Company (HICEC), a subsidiary of HIE in 2004. In 2007 HICEC in turn evolved into Community Energy Scotland (CES)[9] with the role of providing education, finance and practical advice for communities throughout Scotland aiming to achieve a more sustainable energy future.

CES is still thriving today and has been one of the driving forces behind the growth of community renewa-

---

* Highlands and Islands Enterprise, as the name suggests, is the Scottish government's development agency for the north and west of Scotland (the Highlands and Islands). However, it has a larger remit than most development agencies in that it includes community as well as economic development.

ble generation in Scotland. It's estimated that so far it has supported the installation of around 37MW of community-owned renewables, 27MW of electricity generation and 10MW of heat generation. It has more recently also initiated a number of innovative projects where, for example, community-owned wind farms sell power locally at lower rates, thus also beginning to directly address issues of fuel poverty.

## A movement appears

Back in the early 2000s, the success of the community buyout of Gigha and the construction of the wind turbines on the island had a romance and a poetry that were irresistible to the media and consequently rightly received a huge amount of publicity. Despite all the good work of CAT and CSE over the years, and the success of the Baywind Co-operative, it was probably at this point that many ordinary people first became aware of the notion and potential of community energy.

With the level of media coverage of climate change still increasing at this time, the two other pioneering projects that emerged in the first half of the 2000s and received noticeable amounts of publicity were the Fintry Wind Turbine in central Scotland and Ashton Hayes Going Carbon Neutral, Cheshire. As described in Chapter 1, these two projects took very different approaches but both were seeking to engage ordinary people in energy use, carbon emissions and climate change.

Fintry set up a local development trust charity that took a stake in a local commercial wind farm. The aim of this was to generate an income that could be used for other community and domestic energy reduction projects within the village. This would have direct benefit for the people living in the local community, but the aim was also to create a model of community engagement in commercial developments that would reduce the risk of public antipathy to wind farm developments that people often felt were imposed without any real local engagement.

The route that Ashton Hayes Going Carbon Neutral took was very different. Rather than the focus being on energy generation, Ashton Hayes set a target for energy reduction for the community – in this case, of being England's first carbon neutral community – encouraging householders to buy into this target and providing advice, guidance and measurement to help people get there.

**Opening of community turbine, Fintry, Scotland.**
(photo: Peter Skabara)

By the mid-2000s a number of other energy generation projects had started to emerge. For example, small-scale hydro projects were proposed and ultimately built at both Settle in North Yorkshire and Torrs in Derbyshire. Both of these schemes are examples of BenComs, a type of co-operative that is now an extremely popular vehicle for local investment in energy generation schemes and particularly hydros (see page 118). These schemes are also interesting because they're among the first in the UK to use Archimedes screws to drive the turbine. Similarly, a number of other behaviour change, and community energy reduction schemes had been founded along the lines of Ashton Hayes, for example Going Carbon Neutral Stirling.

However, one of the limitations of the ROCs support mechanism was that while it achieved its aim of making larger wind farm renewable energy generation viable, the level of support provided didn't work nearly so well for smaller projects, which typically have a much higher capital cost for each kilowatt of generated electricity. Many people wanting to develop community renewable generation schemes were therefore caught in the middle. Developing a commercial-scale wind farm takes a number of years and hundreds of thousands of pounds and is well beyond the capacity of most communities unless they receive a lot of help. On the other hand, smaller projects of the size that community groups could hope to undertake successfully were very often not financially viable because the income they would generate wouldn't justify the capital costs required.

## FITs, the RHI and the rise of the BenComs

For community electricity generation, this changed in April 2010 with the introduction of the Feed-in Tariffs (FITs). FITs had been used in other countries for some time. One of their characteristics is that they pay generators different amounts for electricity depending on the cost of producing it - so, for example, someone generating electricity from a small wind turbine will be paid more per unit that someone generating electricity from a larger wind turbine. (Exactly how FITs work is described in detail in Chapter 12.)

When they were introduced, FITs were specifically aimed to encourage homeowners and organisations like schools and community groups to generate their own electricity. In the first five years since their introduction in 2010, Ofgem has registered over 2,300 renewable installations as being owned by community groups accounting for over 43MW of installed capacity.[10]

The vast majority of these - over 2,100 (over 30MW of the installed capacity) - are for photovoltaic panels but there are still 150 wind installations that account for 11.7MW of capacity.

| Year | Number of Installations | Capacity |
|-------|------------------------|----------|
| 1 | 386 | 4.79MW |
| 2 | 664 | 8.07MW |
| 3 | 731 | 12.29MW |
| 4 | 270 | 6.17MW |
| 5 | 271 | 11.84MW |
| Total | 2322 | 43.34MW |

Table 1: Feed-in Tariff community installations by year.

The Renewable Heat Incentive (RHI) was announced at the same time as the Feed-in Tariff. It operates in much the same way as the FIT, but applies to heat generated from renewable sources, rather than electricity. Although it was announced at the same time as the FIT, the RHI wasn't introduced until some years later, in November 2011.

Like the FIT, the RHI applies to a range of technologies including biomass (wood-fuelled) boilers, ground-source and air-source heat pumps, and solar water-heating, and pays the generator of the heat different amounts, depending on the cost of producing it.

| Technology | Installations | Total heat supplied (MWh) |
|---|---|---|
| Air-source heat pumps | 21,049 | 251,171 |
| Ground-source heat pumps | 6,955 | 139,617 |
| Biomass | 11,657 | 492,865 |
| Solar thermal | 7,699 | 15,490 |
| Total | 47,370 | 899,144 |

Table 2: Total Domestic RHI installations by May 2016[11]

The arrival of FITs and the RHI resulted in a step change in the development of community energy, allowing communities to undertake generation projects that would previously not have been financially sustainable. Hundreds of community generation projects have appeared over the past few years on the back of FITs and the RHI, including solar panels on schools and village halls, community hydro projects, community wind turbines, biomass boilers replacing oil boilers and solar panels being combined with heat pumps in village halls.

| Technology | Installations | Total heat supplied (MWh) |
|---|---|---|
| Air-source heat pumps | 114 | 4,000 |
| Ground-source heat pumps | 455 | 88,000 |
| Biomass | 13,121 | 7,000,000 |
| Solar thermal | 200 | 3,000 |
| Biogas | 41 | 51,000 |
| Total | 13,731 | 7,146,000 |

Table 3: Total Non-Domestic RHI installations by May 2016[12]

One of the models that has proved to be very popular, particularly for community hydro and wind turbines, is using a legal structure called a Community Benefit Societies, usually just referred to as BenComs. BenComs are described in more detail in Chapter 13 but they work very well for projects that are looking to raise money for local investors in a way that is equitable and that ensures that profits can be retained for the benefit of the community. For community energy, the use of this approach was pioneered by the Settle and Torrs hydro schemes. There are now community energy BenComs throughout the country.

## UK government recognition

In 2014 the UK government published its first Community Energy Strategy report.[13] The report recognised the important role that community energy could play as part of the overall approach to energy in the UK. It's described in more detail in Chapter 4. The single biggest initiative it contained was the arrival of the Urban Community Energy Fund (UCEF), which partnered the already existing Rural Community Energy Fund (RCEF).

By May 2016, UCEF had approved 71 grants totalling over £1 million and RCEF had approved 73 grants totalling over £1.2 million.

It appeared for a while that community energy as a movement was strong and growing: its role recognised by the UK government and by the devolved administrations around the UK; there have been well-attended national and regional conferences held; Community Energy England and Community Energy Wales have joined Community Energy Scotland as bodies representing the movement.

However, in reality, it's still very fragile and occupies a role that's largely on the fringes of government policy. The future is far from secure, particularly when there is a government that is at best lukewarm to the importance of renewable energy generation and demand reduction. The changes to the rules concerning FITs and the RHI since 2015 have killed many potential projects and damaged the movement as a whole.

Chapter 4

# GOVERNMENT POLICY

Last year saw the largest ever growth of community energy projects, not only in terms of numbers but also in scale and the benefits that community energy delivered. At the same time however the sector faced unprecedented changes to policy, finance and regulation, and the model that delivered all those projects is now broken.

"Community energy: The way forward" 10:10.[1]

The policies that have allowed the community energy sector to grow were something of a patchwork, with many of them having very different aims and purposes. While the UK government was very slow to recognise the importance and potential of community energy, that appeared to change when the UK government Community Energy Strategy report was published in 2014.[2]

However, as the quote above from 10:10 (a charity that runs community projects focused on tackling climate change) suggests, recent UK government policies have damaged the sector considerably. Many of these changes weren't aimed specifically at the community energy sector but the cumulative effect of them has been significant.

The most important change was to the Feed-in Tariff, the value of which has been reduced and a cap on the number of installations imposed. Other changes that have had an effect include the withdrawl of tax relief from community energy schemes, the early closure of ROCs for onshore wind and solar PV below 5MW, and an effective moratorium on new wind applications in England.

For many years, the devolved Scottish and Welsh governments have been significantly better than Westminster in recognising the potential of community energy and this remains the case today.

This chapter highlights the main policies and initiatives that the various governments have put in place that have helped community energy projects.

## UK government community energy strategy 2014

As mentioned in the previous chapter, the UK government had published very little and shown very little interest in community energy over the years, but with the growth of the movement and an increasing interest in community energy, the government published its first report in January 2014.[2]

Although a variety of government policies had been key to the growth of community energy, before this report there had been little or no recognition from the UK government for the role of community energy could play. This was even admitted in the ministerial foreword to the report, where Ed Davey, then Secretary of State for Energy and Climate Change stated:

> For too long, community energy has been a policy footnote, with all the focus on big generators and individual households – all but ignoring the potential of communities to play a key role.

The report recognised that community energy could have a significant impact:

> [Community-led action] can often tackle challenges more effectively than government alone, developing solutions to meet local needs, and involving local people. Putting communities in control of the energy they use can have wider benefits such as building stronger communities, creating local jobs, improving health and supporting local economic growth.

It also stated that, "Our ambition is that every community that wants to form an energy group or take forward an energy project should be able to do so, regardless of background or location."

However, it was also clear that there were significant barriers to community energy groups being successful, including those of capability and capacity:

Activities are more likely to succeed where the community has access to the right information, advice and expertise. For community energy to achieve scale, we need to empower groups to learn from each other and share information about what works.

The report recognised that community energy could have wider benefits for society and identified four areas beyond climate change and energy bills:

- Building stronger communities.
- Developing new skills.
- Bringing financial benefits.
- Reducing costs.

The report identified four main areas of activity for community energy:

- Generating electrical or heat energy.
- Energy reduction.
- Managing or balancing energy use.
- Purchasing energy (e.g. collection buying).

This all sounds great but the key question, of course, is what practical measures have taken place since the report was published that can really help community energy projects?

## Urban Community Energy Fund

Possibly the biggest initiative has been the introduction of the Urban Community Energy Fund (UCEF). This is the partner to the RCEF described on page 37 but, as the name suggests, available to urban communities rather than rural. The UCEF wasn't launched until November 2014 and has only £10m for projects compared with the RCEF, which has £15m.

## Shared ownership

It was recognised in the report that shared ownership is one way that communities can have a greater sense of involvement in commercial energy installations. Out of this a Shared Ownership Taskforce was set up, which brings together people who have experience in shared ownership of renewable generation from both the commercial and the community side.

On top of this, the government has passed legislation called the Community Electricity Right, which gives it the right to require commercial renewable electricity developments to offer the opportunity of shared ownership to communities. However, the government has also stated that this is a power that it doesn't want to use and it would much rather that there was a substantial increase in the offer of shared ownership through voluntary rather than compulsory means.

At the time of writing, it's impossible to be sure how this is going to play out. It should be the case that more opportunities are being provided for communities to get involved with commercial projects. However, without the right support the number of communities who have the capabilities and capacities to take advantage of these offers is likely to remain relatively small.

## Community Feed-in Tariff

The maximum size of an electricity generation project that is eligible for the Feed-in Tariff is 5MW. The report recognised that one consequence of this was for community groups to limit their projects to 5MW because of the barriers involved in going to the next step of the Renewable Obligation Certificate (ROC) or Contract For Difference (CFD) markets (see page 54). The report therefore proposed to increase the maximum size for eligibility to 10MW for community-owned projects. This was described as the "Community FIT".

Unfortunately, although various useful tweaks were made to the rules for FITs that affect community groups following the publication of the report, the government didn't go ahead with the "Community FIT". The reasons were that a change of this type would require state aid approval from the European Commission and it would be difficult to achieve this because there wasn't enough evidence that the costs of community projects are higher or that there is a genuine market failure.

## Other initiatives and proposals

As well as the specific points mentioned above, there were a number of other initiatives:

- Advising local authorities in England and Wales that they should give more recognition to the positive benefits that community energy can provide.

- Convening the Powering Up conference that took place in Oxford in September 2014 and was followed by Powering Up North that took place in Manchester in February 2015. Both conferences were viewed as being significant successes with over 200 delegates.

- Committing £430,000 to funding a new Green Open Homes national network.

- Setting up a "One-stop Shop" information resource for community energy groups that will include a range of services and resources that will help build the capability and capacity of community energy. (This was supposed to be set up in 2014 but at the time of writing it hasn't yet happened.)

- A £100,000 community energy saving competition.

- A package of Community Energy Advice pilots to identify the most effective community-based approaches to cutting waste and spending less on energy through behaviour change, including a £500,000 scheme to trial and scale up peer-to-peer approaches to energy saving advice in housing associations.

- Extending the Green Deal Communities scheme funding from £20m to £80m.

This last point about Green Deal Communities sounds wonderful for community energy when seen as a headline, but is actually a scheme aimed at local authorities to do street-by-street Green Deal work. In itself, there's nothing wrong with it but unfortunately it doesn't have much to do with community energy.

## The FIT and RHI

The UK government's Community Energy Strategy report and the initiatives it included were welcome, but community energy existed and was growing strongly before the report was produced. The support mechanisms of the FIT, the RHI, and before them, to a lesser extent, the ROC were major factors in this growth.

FITs were announced by Ed Milliband in October 2008 and were introduced in April 2010. Although the RHI was announced at the same time, its introduction was much delayed, with the Non-Domestic RHI being introduced in November 2011 and the Domestic RHI in April 2014.

These support mechanisms made renewable energy generation at the scale that's viable for communities financially sustainable and although they were primarily intended for householders or businesses, they were equally applicable to enterprising community groups.

However, they also cause problems for community groups:

- They require the group to source capital to build the generation facility and once it's been built pass on a great deal of the income generated to the providers of that capital.

- On the basis that some technologies are expected to get cheaper as volumes build, so the government uses a system called degression to regularly reduce the level of support.

Degression introduces an uncertainty into schemes that can often take a year or more to reach the point where they start generating electricity and producing an income.

Obviously any project that takes some time to reach fruition is at risk because the income it's intended to generate might well be significantly reduced by the time it reaches the stage when it can be installed. This is a problem for any renewable energy project, whether it's community-based or commercial, but the problem is felt particularly strongly by community groups which often move relatively slowly because it's volunteers who are driving the projects and they have only limited time.

Although degression was introduced with the FIT as a way of stopping profiteering as the price of renewable technologies was expected to drop, it's increasingly seen as the method that is used by a government that's not particularly keen on renewables to ration the number of renewable installations.

This approach by the UK government became more obvious when various announcements about FITs were made in 2015 following the UK general election. From January 2016, FIT rates were significantly reduced. For example, before the reduction, a typical domestic PV installation with a capacity of less than 4kW would receive a tariff of between 11.67p and 12.88p per kilowatt hour of electricity produced. The FIT is now set at 4.39p per kilowatt hour.

On top of this, quarterly budgets have been set for installations of all technologies. If a budget is exceeded during a quarter, FIT applications are queued for the next quarter.

## Local Energy Assessment Fund

The Local Energy Assessment Fund (LEAF) was aimed at not-for-profit groups and designed to help them assess how they could improve energy efficiency in their local area, increase renewable energy generation, and benefit from policies such as the Green Deal and the RHI.

Around 200 groups benefited from awards of around £50,000 to cover the costs of work such as:

- Analysis of potential for energy efficiency and renewable energy measures (e.g. street-by-street home energy audits).

- Community engagement to highlight these opportunities.

- Demonstrations of technologies in local buildings (e.g. installing and showcasing solid-wall insulation).

- Feasibility studies for community-scale renewable energy schemes.

## Rural Community Energy Fund

The Rural Community Energy Fund (RCEF) provides community groups in rural areas of England with access to finance to allow them to explore electricity and heat generation projects.

How the fund works is described in more detail in the Chapter 14 but broadly speaking it provides grants for initial feasibility work on projects of up to £20,000 and loans of up to £130,000 for pre-planning activities. The loans are "contingent", meaning that if the project doesn't go ahead then the loan is written off, but if the project does go ahead then it needs to be repaid with a 45% premium. The RCEF only operates in England (there are similar schemes in other parts of the UK, which are described below) and has a total fund of £15m.

This approach was based on the success of the Scottish government's Community And Renewable Energy Scheme (CARES) scheme (see page 40) and addresses one of the main stumbling blocks for community energy generation projects where significant sums of money are needed to prove the feasibility of a project and take it through the design and planning process. Few community groups had or could access enough money to take projects through this stage because the money was always "at risk" – if the project was proved not to be feasible or if, for example, planning consent was refused the money, could not be recovered.

The launch of the RCEF was in June 2013. In its first two years of operation it had allocated over 70 awards, totalling over £1.2million.

# Energy Company Obligation (ECO)

Under the Energy Company Obligation (or ECO) the big six energy suppliers are required to help householders save on their energy bills and carbon emissions. There are three strands to ECO:

- The Carbon Saving Community Obligation, which provides insulation measures to

- households in specified areas of low income. It also makes sure that 15% of each supplier's obligation is used to upgrade more hard-to-reach low-income households in rural areas.

- The Affordable Warmth Obligation, which provides heating and insulation measures to homeowners that receive particular means-tested benefits. Its aim is to support those on low incomes that are vulnerable to the impact of living in cold homes, including the elderly, disabled and families.

- The Carbon Saving Obligation, which covers the installation of measures like solid-wall and hard-to-treat cavity wall insulation.

ECO was introduced in 2012 and replaced other similar obligations on the energy supply companies like the Carbon Emissions Reduction Target (CERT) and the Community Energy Saving Programme (CESP).

ECO appears to have had a significant impact, with over 900,000 measures being installed by August 2014,[3] the latest date there are any figures for. It's also been possible for community groups to engage with ECO as well. For example, funding for biomass woodchip boilers for district heating schemes has been available, provided the beneficiaries of the scheme meet the criteria.

ECO was initially due to end in March 2015 but following a consultation on its future the scheme has been tweaked and extended by two years to March 2017.

## Investment incentives

There are a number of "investment incentives" that have been relevant to community renewable schemes. The key schemes were the Enterprise Investment Scheme (EIS) and the Seed Enterprise Investment Scheme (SEIS). Both of these schemes allowed investors in small, relatively high-risk companies to claim tax back on part of their investments. Share offers for investments supported by the FIT or the RHI weren't eligible for these schemes. The exception to this was if they were recognised as community energy schemes.

These tax breaks made investment in community energy co-operatives or BenComs much more attractive than they would otherwise have been and were a key factor in the growth of schemes funded in this way. However, in 2014 the UK government announced that community energy schemes would no longer be eligible for EIS or SEIS but they would extend another similar scheme the Social Investment Tax Relief (SITR), to include community energy. However, in a further announcement, the government then reversed this and stated that community energy schemes will be excluded from Social Investment Tax Relief (SITR). This means that there is no tax relief scheme that is applicable to community renewable generation projects.

## The Green Deal

The Green Deal scheme launched in January 2013 was supposed to be the "biggest home improvement programme since the Second World War". It was aimed primarily at householders and provides an energy efficiency assessment together with suggestions for improvements. There is also a government-backed finance scheme for making improvements, which are paid back by the householder through charges being added to their electricity bills.

It was stated that this was the scheme that would tackle energy reduction but was widely criticised as being cumbersome, poorly advertised and providing a process that householders found difficult to work through successfully.

In July 2015 it was announced that the scheme would close. By this time, over 600,000 Green Deal assessments had been undertaken but only 15,000 plans were "live" with all measures installed.[4]

Because of the way it's been set up, there's been very little that community energy organisations have been able to do around the Green Deal. This is undoubtedly an opportunity missed: community groups have an enthusiasm for energy reduction measures and an understanding of local neighbourhoods. With a scheme structured in the right way, local neighbourhood energy reduction schemes driven by community organisations could have made a big difference to the number of houses benefiting from having major improvements made to their energy efficiency with obvious benefits for the householders.

## Scottish government

In contrast to the UK government, the Scottish government has recognised community energy for over 10 years, with the Scottish Community and Householder Renewables Initiative (SCHRI) being launched in 2002 with grants of up to £100,000 for community projects and a network of development officers providing advice and support.

More recently the Scottish government set a target of 500MW of community and locally owned renewables by 2020. This target seems to be well within reach, with the latest total figure from June 2014 being 361MW in operation of which 46MW was community owned (up from 285MW in total and 43MW community-owned, a year earlier).

Over the years the Scottish government has launched a number of initiatives to support community energy. The most recent publication on community energy is a policy statement from September 2015.[5] As well as indicating continued support for the existing initiatives supporting community energy, this statement introduced the notion of Local Energy Economies as being a direction of travel for the future:

> The challenge is to encourage Scotland's communities to grasp the opportunities of a whole system approach to community energy. Low carbon energy systems can involve the integrations of a range of technologies: renewable energy generation – coupled with energy storage, the use of waste heat and community heat systems, smart grids and demand reduction measures, improvements to energy efficiency.

## Practical measures

The practical measures provided to community groups by the Scottish Government start with a range of different types of financial support.

### Community And Renewable Energy Scheme (CARES)

The first Community and Renewable Energy Scheme was launched in 2009. Like the RCEF and UCEF mentioned earlier, its role is to support the early stages of renewable energy generation schemes. It has two main funding strands:

- Providing grants of up to £10,000 for feasibility studies and community consultants.

- Providing loans of up to £150,000 for pre-planning activities.

Like the RCEF and UCEF, the loans are written off if the project can't go ahead. Unlike the RCEF and UCEF, the loans don't have to be paid back at financial close but they do have an annual interest rate of 10% on any money borrowed. CARES is available to rural businesses as well as community groups and by June 2014 a total of 133 loans had been offered under the scheme with a total value of £12.5m.

As well as providing grant and loan funding, CARES also provides practical support and advice for community groups through a network of local development staff. Over the years since it was first launched, CARES has been constantly evolving with the latest initiative being the Local Energy Challenge Fund (LECF).

### Local Energy Challenge Fund (LECF)

The Local Energy Challenge Fund (LECF) was initially launched in 2014, and has been created to "demonstrate the value and benefit of local low carbon energy economies".

It provides grant funding for innovative projects that link local energy generation to local energy use. It operates in two phases, with the first phase providing support up to £25,000 for feasibility and developing project proposals. A number of phase 1 projects are then provided with support for phase 2. In 2015 around £21 million was awarded to the first tranche of six successful phase 2 projects.

The LECF is open to a wide range of organisations. As well as community groups, this includes local authorities, social landlords, universities and commercial organisations.

### Renewable Energy Investment Fund (REIF)

More recently, the Scottish government launched the Renewable Energy Investment Fund (REIF) scheme. Where CARES provides support for pre-planning activities for projects, REIF provides financial support in the form of loans for post-planning activities. (Like CARES, REIF has a broader remit than just community renewables, covering among other things marine renewables and renewable district heating.)

REIF is designed to "fill the gap" where viable community energy projects aren't able to complete funding of projects i.e. where there is market failure. It provides loans, guarantees and equity at market rates and is managed by the Scottish Investment Bank.

Since it launched in 2012, REIF has invested over £3m in community energy projects. Unlike other funds such as CARES, REIF doesn't work in a prescribed fashion but provides a flexible lending service that is tailored to individual projects that have funding gaps.

### Climate Challenge Fund (CCF)

The Climate Challenge Fund (CCF) launched in 2008 and has a much wider remit than simply energy. It provides "funding for community groups that want to tackle climate change through community-led projects". Among the projects it has provided funding for are the refurbishment of community-owned buildings, local energy efficiency advice, waste recycling, active travel projects and local food growing.

Since launched, it has awarded over 650 grants to nearly 500 communities, and support is scheduled to have reached £80 million by 2016. The CCF specifically doesn't provide funding for the types of things that the CARES, REIF or LECF provide.

The CCF is open to projects that are led by the community and are not-for-profit. All projects must lead to carbon reductions, and projects are expected to leave a sustainable legacy once they have completed.

The final round of CCF projects were expected to end in March 2016. However this has been extended and the next set of projects is expected to run through to March 2018. The future of the fund beyond this is not yet known.

### Others

There are a number of other funds that the Scottish government runs which, while not being specifically for community energy projects or community energy groups, are relevant. Among them are the People and Communities Fund which "supports community anchor organisations to grow and strengthen by delivering outcomes to meet the aspirations of their communities";

and the Scottish Land Fund which awards grants to help rural communities buy land, including sites for the development of renewable energy opportunities.

### Community Energy Scotland and Local Energy Scotland

Scotland is blessed with not one, but two organisations that provide support for communities wanting to undertake energy projects.

Community Energy Scotland (CES) began as the Community Energy Unit of Highland and Islands Enterprise (HIE), established in 2002. CES has had a key role in the growth of the sector in Scotland, particularly in the Highlands and Islands areas. When CARES was first launched, it was CES who managed and delivered the scheme.

A membership organisation, CES now operates across the whole of Scotland and has over 330 community groups who are members. It provides practical help on renewable energy development and energy conservation projects, and runs a well-attended annual community energy conference.

Local Energy Scotland (LES) is a consortium of organisations – the Energy Saving Trust (EST), Changeworks, The Energy Agency, Save Cash and Reduce Fuel (Scarf) and The Wise Group. It was formed to deliver the various CARES initiatives and took over the administration and management of CARES from CES in 2013. It has a remit wider than just community energy, offering advice to rural businesses as well. It has a network of local advisors and sees its role as providing:

- Advice and support to help communities and rural businesses develop renewable energy schemes.

- Advice on funding streams.

- Support to access CARES.

- Support to access REIF.

# Welsh government

The Welsh Government has been broadly supportive of community energy for a number of years. The most significant initiative that has resulted from this support is the Ynni'r Fro Community Programme. The scheme uses European Structural Funds to provide social enterprises grant aid, loans and advice to help develop community-scale renewable energy projects in Wales.

The scheme was launched in 2010. It provides up to £30,000 for pre-planning activities like environmental surveys and planning applications but it can also provide loans of up to £300,000 or grants up to £250,000 towards the capital costs of renewable energy projects. On top of this, there is also a network for technical development officers across Wales to help community groups with their projects.

# Northern Ireland Executive

The most significant difference in policy in Northern Ireland is that there is currently no Feed-in Tariff for electricity generation projects. Instead Northern Ireland has what's called the Northern Ireland Renewable

Obligation (NIRO). In reality, although the mechanism looks quite different, the end result for renewable electricity projects is very similar to the FIT, with generators paid an amount based on the technology and capacity of the scheme plus a fixed amount for every unit exported. NIRO is due to be phased out in 2017 and replaced by a FIT.

Northern Ireland also has a slightly different approach to the Renewable Heat Incentive as well. Non-domestic renewable heat installations are part of the UK-wide RHI and administered by Ofgem. However, the domestic side of the scheme works a bit differently, with installations receiving an up-front grant (which varies based on the technology) and ongoing payments (also based on the technology).

There is currently no equivalent of RCEF, UCEF or CARES in Northern Ireland and no funded support organisation either.

## COP21

The Paris Agreement was adopted in December 2015 at COP21, the 2015 Paris Climate Change Conference. At the time it was signed there was hope that it would lead to new policies in the UK addressing climate change and that some of these policies would boost community energy. Over a year after the deal was signed there is no policy evidence that the current UK government is any more interested in community energy than it was previously.

# PART 2
# THE TECHNOLOGIES

Part 2 provides descriptions of each of the main technologies that are used in most community energy projects together with short case studies showing how they work in practice. As described in Part 1, sustainability is a key element of community energy and therefore the technologies used are all forms of renewable energy.

As well as describing the technologies, this part of the book also explains the workings of the support mechanisms that the UK government put in place for renewable energy generation – the Feed-in Tariff (FIT) for electricity generation and the Renewable Heat Incentive (RHI) for heat generation.

The main technologies described are:

- Solar thermal and solar photovoltaic.

- Wind turbines.

- Hydropower.

- Heat pumps.

- Biomass.

- Anaerobic digestion (AD).

There is also a chapter on projects that don't involve energy generation at all but are based on the important subject of community energy reduction, covering things like home energy audits, domestic retrofits and behaviour change.

Much of the content in this part of the book was taken or adapted with permission from material produced by the Centre for Sustainable Energy, in particular from PlanLoCaL (www.planlocal.org.uk) and I'd like to thank CSE for allowing me to use their material.

Chapter 5

# SOLAR

Solar installations are very often part of community energy projects, in particular projects that are refurbishing or enhancing existing community buildings. Generating heat or electricity from solar can make buildings viable that wouldn't have been otherwise.

Solar power takes energy from the sun and, depending on the type of solar panel, uses this energy to heat water or generate electricity.

Both types of solar panel work best when they face south – this way they get most sunlight. They'll work OK, though not as well, if they face south-east and south-west, but outside this range they will probably not be cost-effective. In the UK, they also need to be tilted at an angle of about 35° (about the same as the average pitched roof) in order to catch the most sun, although this varies depending how for north or south you are. If the panels are installed on a flat roof – or even standing on the ground – they are placed on frames, which tilt them at the optimum angle.

Solar panels obviously don't like shade – particularly if they're the electricity-generating type. Even slight shadows cast by nearby buildings or trees can make them ineffective. They will produce more heat or electricity during the summer than in the winter and they are also more effective the further south in the UK you are.

## Solar thermal – how does it work?

The type of solar system that heats water is known as solar thermal, and works very well for the hot water demands of homes, community centres, schools and swimming pools. It's especially suitable for venues with high usage during the summer months, such as outdoor pools and cricket pavilions.

**Solar thermal panels**
(photo: iStockphoto/ricochet64)

Solar thermal panels are essentially shallow boxes with glass (or clear plastic) covers. They are insulated from behind and designed to absorb the heat of the sun like a highly efficient mini-greenhouse. Inside the panel is a network of pipes through which a mixture of water and antifreeze passes. On a very sunny day this liquid can get very hot indeed – even reaching boiling point.

The hot liquid is pumped from the panel into a heat-exchanger that is coiled inside a hot water tank, which stores the water you're going to actually use. The heated-up mixture of water and antifreeze never comes into contact with the water in the tank. But it does cause it to heat up, and by so doing provides hot water for showers, baths, washing up and other purposes. Solar thermal is not usually suitable for radiators.

Because they get very hot, all solar thermal systems need to be fitted with some sort of venting mechanism to allow for expansion, and a safety system to prevent scalding. Reputable installers will ensure both of these are done, but don't be afraid to check that they've considered this and ask them to explain how it'll be done.

Because you don't necessarily want to use all this hot water the very instant it is heated, all solar thermal systems must have a hot water storage tank. If you already have a hot water tank, you can either have an extra cylinder that works in conjunction with it, or you can have an entirely new cylinder fitted. Single cylinder systems are taller and thinner than normal immersion tanks to allow for stratification* of the water. But they do take up less space overall than having two separate tanks.

## Types of solar thermal

There are two types of solar thermal panels: flat plate collectors and evacuated tube collectors. The latter are more efficient, but are also costlier and less robust. Evacuated tube collectors contain a series of glass tubes through which the pipework containing the water and antifreeze mixture runs. These tubes have a vacuum inside, which means the temperature can get considerably higher than in a flat plate collector. It is also possible to rotate each tube in order to maximise the amount of sunlight they catch. This is particularly useful if your roof's orientation isn't quite due-south.

## Solar thermal and earning money from the Renewable Heat Incentive

As a technology that produces renewable heat, solar thermal is eligible for the UK government's Renewable Heat Incentive (RHI).

---

* It's important for the occupants of the building that hot water is available at all times so there must be hot water in the cylinder. However this presents a problem for a solar system because solar heating is unpredictable, and if the cylinder is full of hot water the solar energy would be wasted. Solar cylinders therefore have a heat exchanger at the top for a conventional boiler, and one at the bottom for the solar loop. The boiler ensures that the water at the top of the cylinder is hot at all times and because cold water is denser than hot, this hot water will float on top of the colder water. When solar heating is available during the day this will then heat the cold water at the bottom of the tank.

As described in Chapter 4, the RHI is a mechanism introduced by the government to encourage the installation of heat-generating renewable technologies. Chapter 12 describes in detail how the RHI works in practice and what the requirements are for solar thermal installations.

### Solar thermal case study

Bovey Tracey is a small town sitting on the edge of Dartmoor, Devon which is lucky enough to have its own community swimming pool.[1] This pool is well used, with around 19,000 swimmer entries per year. When the swimming pool association decided to undertake a major rebuild, they felt it would be sensible to look at their sustainability and running costs at the same time.

To help them, they enrolled Bovey Climate Action (BCA) – a local group with plenty of experience in this area. BCA's initial feelings were to cut costs by using solar panels to heat the pool. But the pool has to be kept at a temperature of 30°C, and to do this they would have needed enough panels to cover at least twice as much roof space than was available to them.

BCA decided a full energy audit of the facility was necessary, and the Devon Association of Renewable Energy was recruited to undertake this. The audit had a number of recommendations, including installing a biomass system to heat the pool – but because the pool is currently heated using mains gas, such a system probably would have meant increasing the pool's running costs, not reducing.

Undeterred, the association opted for a plan to just heat the water for the showers by installing solar thermal panels. The response from users to the plan was very positive, and so funding of around £8,000 was obtained to proceed.

Getting the money was the biggest challenge, with plenty of forms to fill out. The total cost was around £8,000. Most of this, around £5,000, came from the Dartmoor Sustainable Development Fund, with the rest coming from the swimming pool itself, BCA and other small grants. Installation, though, was straightforward; with the local community undertaking most of the work. Day one was spent on the plumbing, and days two and three were spent installing the panels.

The panels work by preheating the cold water, which is then fed to the gas boiler, which supplies the showers. The panels use the more efficient "evacuated tube collectors" and there are times when the gas boiler isn't required at all.

Part of the grant conditions also required the pool to install a panel to show users how much energy has been produced. This now sits proudly by the entrance. A full analysis has yet to determine how much money and $CO_2$ the pool has saved since the panels were installed, but BCA and the pool association are hopeful that, with their proposed biomass or additional solar developments, and the introduction of the RHI, it will soon be able to save even more.

# Solar photovoltaics – how does it work?

The type of solar energy that generates electricity is called solar photovoltaic, or solar "PV". Photovoltaic is a combination of the Greek word for light (*photos*) and the word *volt*, and literally means "the generation of electricity from light".

PV panels or "modules" are made up of a set of solar cells that are connected to each other in series i.e. in a single line. When light falls on a cell, electricity is produced. This electricity is then fed into an inverter which changes the power from direct current to alternating current, which is the kind of electricity used by appliances like fridges or TVs.

**Fitting solar photovoltaic panels.**
(photo: iStock.com/compuinfoto)

The amount of electricity produced is determined by the solar cell that receives the least sunlight. This is important because it means that if just one cell is in the shadow of a building or even just a lamppost, this can significantly reduce the production of electricity. In theory, there are ways to wire up the system to mitigate this effect, but this throws up design barriers and costs would be higher.

So the main point to remember is that you want your panels in a place where there will be as little shade as

possible. Be careful about seasonal changes in the sun's height, too. The sun is lower in the winter and therefore a tree that doesn't cast shade in June could be a problem in December.

## Types of solar PV panel

There are three main types of PV panels – monocrystalline, polycrystalline and amorphous/thin film.

Monocrystalline panels are made of cells of wafer-thin slices of silicon crystal, characterised by their distinctive pattern of small bevelled-edge squares that look black or dark blue. These are the most efficient panels available, but also the most expensive. They're also quite fragile – something to bear in mind on a public building where vandalism might be an issue.

Polycrystalline panels are made of cells that are cut from an ingot of melted and recrystallised silicon – this gives them a very distinctive crackled "shimmer". They're less efficient than monocrystalline, but a little cheaper.

Both crystalline types of panel can be incorporated into glass or glass laminates, so that they can be part of a window and offer partial shading. However, it is important to note that they also get quite warm and could cause the room below to overheat.

Amorphous or thin-film panels can be bent to fit the architecture on which they are mounted, and as they are much more robust than other panels, they may be a good choice if vandalism is a potential issue. They're much cheaper than crystalline panels, but only a third as efficient – so you need more of them to get the same output. However, they do perform better on slightly cloudy days (of which the UK has many) or if partly shaded. Their relative cheapness, robustness and good performance in cloudy conditions can make them a very sensible choice – especially if you've got lots of suitable roof space.

Solar PV systems don't just have to be panels mounted to the roof – although these bolt-on systems are the cheapest. PV tiles and slates are also available but they're more expensive. If fitted when the roof needed replacing anyway, you can offset the cost of new ordinary tiles. And they may be more acceptable on listed buildings such as churches.

## Earning income from the Feed-in Tariff

The Feed-in Tariff (FIT) is a government scheme that rewards people or groups who generate renewable electricity. Unlike the Renewable Heat Incentive (RHI), there is just one FIT scheme that covers both installations on houses and any other locations.

There are a number of different ways this can work. If your group owns a community building then the electricity generated by the installation will reduce the cost of your electricity bills and you can also receive money from the FIT.

If you don't own the building then there are still ways you can benefit financially by agreeing to share the benefits with the owner. For example, your group purchases solar panels, which you install on the roof or roofs of local homes, commercial premises or other buildings. The householder or business whose roof is being used benefits from cheap or free electricity – this is their "rent" if you like. Meanwhile the community reaps

the FIT – which can be worth thousands of pounds a year, guaranteed for 25 years. This is index-linked, so it could be used to invest in other projects.

This kind of arrangement should not be entered into lightly though. For example, you will need to have a contract between yourselves and the building owner defining the rights and responsibilities of both parties.

Finally, if you do own a building but don't have the capital required to install solar PV, you can do it the other way round and rent out your own roof space to a commercial company. Again though, there are pitfalls and you need to be very clear who is doing what and who has responsibility for what.

Chapter 12 describes in detail how the FIT works in practice and what the requirements are for solar pv installations.

## Solar PV case study

Matthew Arnold School on the outskirts of Oxford installed 100kW of solar PV in 2010 in two arrays, one of 52kW on the main school building and the remainder on the science building. The panels are all monocrystalline and are roof-mounted, facing south-south-west. At the time of installation, it was the largest solar PV installation on a school in the UK.

The project was undertaken by West Oxford Community Renewables (WOCoRe) a co-operative that builds community-owned renewable energy schemes in west Oxford to generate green energy and provide income to fund carbon reduction projects in the local community. Although WOCoRe managed the project, because it was such a big project and with funding from the Low Carbon Communities Challenge the installation was actually performed by a local installer.

The total electricity generated is around 85,000 kilowatt-hours (kWh) per year and the school uses nearly all of this, with electricity being exported on only a few days in July and August. This means that the school is able to save somewhere around £10,000 on electricity costs per year. The annual carbon savings are around 45 tonnes of $CO_2e$.

At the time of the installation, the FIT for solar PV projects of this size was 31.4p/kWh. The income earned by WOCoRe from operating the panels contributes to an overall anticipated surplus of £25,000 per annum for the organisation, which will be donated to local environmental projects and help create further carbon cuts. The total capital cost for the project was around £300,000.

## Setting up a solar energy project

If you're thinking of setting up a solar energy project, the good news is that, for domestic or other single-building systems, solar panels are probably the most straightforward of all renewable energy technologies. This is the case for both types of solar: hot water and PV.

That said, solar is the renewable technology that suffers most from "cowboy" operators. But weeding out the bad guys is relatively straightforward, particularly if you only contact installers that are registered with the Microgeneration Certification Scheme (MCS).[2] This is a government-supported initiative designed to protect consumers. If you want to benefit from the FIT or the

**Solar panels on Matthew Arnold School in Oxford.**
(photo: Lois Muddiman)

RHI, it's a requirement that the installation is installed by an MCS installer and that the installation has an MCS certificate.

Once you've identified some potential contractors, it's time to do a bit more research. Reputable operators will almost certainly have websites. Check them out, investigate poor reviews and take up references from previous customers.

And, as with other building work of a similar scale, you should get at least three quotes from different installers. As a guide, an average domestic solar PV system currently costs in the region of £1,800 per kW.

### Bulk purchase

If your solar project involves a lot of people in your community all installing solar systems, it should be possible to negotiate a bulk-buy discount. You could also consider setting up a co-operative to purchase solar panels and associated kit, from which members can buy at a discount.

## Planning permission

Whether you need planning permission for solar panels is not straightforward and depends on many factors, including whether the building is a house or not, how big the installation is and whether the building is in a conservation area. A reputable, thorough installer should be able to guide you through the planning permission process and if you have any doubts about the right thing to do, you should contact your local council to ask them for advice on your particular circumstances.

In most cases of installing solar panels on houses, you do not need planning permission or a planning application, as they are most likely considered "permitted development". However, there are a few limitations and conditions, which must be adhered to before you can install your solar panels. If you are not 100% sure, you should contact your local authority and ask them whether planning permission is required for solar panels with regard to your specific situation.

With flats and commercial buildings you'll probably need to seek permission, but unless the building is listed or in a conservation area, this is likely to be granted – particularly if they're relatively discreet.

Also, bear in mind that even if you don't need planning permission, you will need to ensure the installation complies with all the relevant Building Regulations. For example, you need to be sure that the roof structure is strong enough to cope with any additional weight.

Chapter 6

# WIND TURBINES

The wind resource in the UK, and particularly Scotland, is the best in Europe, both onshore and offshore, so in theory there are plenty of opportunities to site wind turbines.

Wind turbines work by converting a portion of the energy in wind into rotational motion. This is then converted into electricity by a generator.

It may sound obvious but, to be economically viable, wind turbines need to be sited in windy places. So what counts as windy enough? The annual average wind speed at hub height (the centre of the rotating blades) needs to be around 6 metres per second (m/s). Commercial developers usually demand speeds in excess of 6.5m/s.

Because the higher up you are the windier it is, wind developers prefer the tallest hub height possible. And it obviously helps if turbines are sited on hills and in open spaces. This is why there can be some heated debate about landscape impact.

What's more, the wind needs to blow from the same direction as much as possible, and be uninterrupted by nearby obstacles such as buildings, high hedges and trees. Turbulent sites are not good, even if the gusts of wind are quite fast.

It's not unheard of for planning or landscape officers to request that a proposed turbine be made lower, moved nearer to the edge of a woodland, or moved further down a hill in order to reduce the landscape impact. The result can be to reduce the wind speed, make it more turbulent and often make the whole project unviable!

## Size matters

Without getting too technical here, it's important that you understand why scale is important. A small wind turbine, 10m tall at the hub, can generate enough electricity for about three average homes. A large turbine, 80m at the hub, can generate enough for 1,200 homes. It seems odd that a turbine only eight times bigger can produce that much more electricity,

Here's why. As we know, the higher you are, the windier it is, so the taller turbine can catch more wind. And what's more, there's a "cubic relationship" between wind speed and power output – so if the wind speed doubles, the amount of electricity produced is multiplied by eight! Secondly, the power output is proportional to the swept area of the turbine – therefore, if you double the length of the blade, power output is quadrupled.

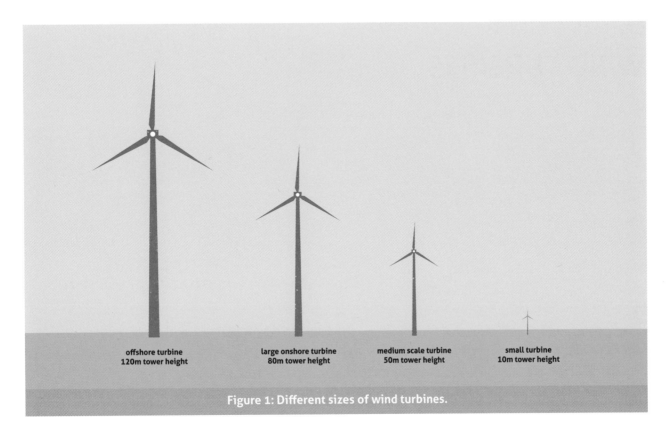

offshore turbine
120m tower height

large onshore turbine
80m tower height

medium scale turbine
50m tower height

small turbine
10m tower height

**Figure 1: Different sizes of wind turbines.**

## What kind of turbines are there?

You will hear people talk of micro-, small-, medium- and large-scale wind. Let's look at the differences:

### Small turbines

Small turbines are usually single machines supplying electricity to specific buildings such as farms and schools. They provide about the same electricity over a year as would be used by 2–5 average-sized houses. Size-wise they're around 10m tall although some are smaller and some larger.

### Medium-scale turbines

With the introduction of the FIT, medium-scale turbines, which had only been marginally viable in the past, became realistic propositions. These are most often installed as single turbines and are typically anything up to 500kW in output and up to a maximum height of around 60m.

### Large-scale turbines

Individual large-scale turbines can also be deployed as single machines, but are often used in groups to form part of a larger planning application for a large-scale wind farm. Wind farms tend to be located in more remote and rural areas. A "large turbine" would be anything over 500kW capacity, up to around 3.5MW. At their largest they could have a tower of 90m, and could provide the same amount of electricity as that consumed by 1,200 houses over a year.

### Offshore turbines

These are the largest of all, and are being used more and more at sea. However, offshore wind farming remains technologically challenging and expensive, due to the hostile environment. Some people think that all wind turbines should be put out at sea, but in the long term we will need to utilise all resources, both offshore and onshore, to meet the challenge of weaning ourselves off fossil fuels.

### Vertical-axis turbines

These use a range of designs in which the rotor shaft spins on a vertical axis. A key difference when compared with horizontal axis turbines is that they operate independently to wind direction – in other words they don't need to be aligned with the prevailing wind and are thus suitable in gusty locations like towns and cities. As a consequence, a few small-scale vertical-axis turbines can now be seen in urban areas, but vertical-axis turbines don't perform well at a larger scale so it's unlikely that you will see big ones in the countryside in the near future.

## Myths and misconceptions

Wind energy, perhaps more than any other renewable technology, is beset by misinformation, and there are a number of myths and misconceptions about medium and large-scale wind turbines doing the rounds. These include that they kill large numbers of birds, don't generate as much power as is claimed, suppress house prices and cause epilepsy.

Hopefully, this introduction has shown you that there are different types of turbine you could consider, and what would make a good site to investigate initially. If you think that wind is something your community will investigate more thoroughly, the section on setting up a wind project will give you more detail.

## Wind, FITs, ROCs and the CFD

The position with regard to financial support for wind projects can be a bit confusing because there were a number of different mechanisms, each with its own acronym. However, in reality, most community wind energy projects will be eligible for the simplest, the Feed In Tariff (FIT).

Larger projects were eligible for the Contract For Difference (CFD), which replaced the previous support mechanism called the Renewable Obligation Certificate (ROC). However, in 2015 the UK government appeared to announce that on-shore wind farms would no longer be eligible for a CFD meaning there was no longer any support mechanism for these types of projects. Very few

100% community-owned projects would have been at this scale, but if the government maintains this policy, it will mean that there will be no more large projects in which communities could have owned a share.

How FITs work and more information about the CFD and ROC can be found in Chapter 12.

# A wind project

If you're hoping to undertake a medium-scale or large-scale wind project, you'll almost certainly need input from a professional consultant at some stage.

However, before you get to this stage, you may want to make some initial investigations yourself of the site you have in mind for your wind turbine. You can do this with a simple map survey and a visit to the site and surrounding area. It might be a good idea to do this before you consult the community – if the site's not viable, there's no point getting hopes up.

Alternatively, you could run some initial consultation events, introducing the community to the idea of renewable energy projects and getting a sense of which technologies people like the idea of. If wind energy isn't popular you may have to think again – although most people will support an initial feasibility study once presented with the facts about wind.

## A DIY site assessment

If you are going to investigate wind, these are some of the questions that you or your community should ask, and for which you don't need to pay a consultant.

1.  What is the approximate average annual wind speed at your prospective site?

    You can find this out online using the DTI wind speed database on the website of RenewableUK, formerly known as the British Wind Energy Association (BWEA).[1] This is a rough guide only, but if it shows 6 metres per second (m/s) or more at 45m above ground level, then it's worth investigating further.

2.  Is the site clear of trees or buildings?

    Remember you want "clean" wind as far as possible but some woodland or other feature is probably OK if you are looking at a turbine that will be significantly taller than any obstacles.

3.  Is it in a national park or AONB?

    This won't absolutely prevent you erecting a turbine, but it will make obtaining planning permission less likely.

4.  Are there any domestic dwellings within 400m of the proposed site?

    If so, your wind project is probably not feasible because of planning restrictions that relate to noise and other impacts.

5.  Are there any bridleways within 200m?

    Turbines cannot be closer than this.

6.  Are there any overhead power cables within falling distance of the turbine?

    Falling distance is deemed to be the height of the turbine's hub, plus the length of the rotor plus another 30m for the sake of safety.

7. Are there any airfields, TV transmitters, or air traffic control towers within 3 miles?

Again, not necessarily a deal-breaker, but you need to be aware of them and possibly take them into account in the design of the site. You will also need to consult with the people who own or manage these facilities.

8. Where is the nearest place you could connect into the national grid?

You obviously need to do this to export the electricity you generate. If the nearest substation is several miles away, the costs of transmission could be too high for all but the larger schemes.

9. How easy is the site to access with large machinery?

Large lorries will need to get up there with the turbine components. Will a new road access need building?

10. Who owns the land?

Some farmers or landowners will welcome an approach to develop a wind farm on their land.

So, you've checked the wind speed, there are no houses, bridleways, airfields or overhead lines too close. You know there's a substation nearby and an access road, and you're confident that the landowner can be contacted. It's probably time to talk to a consultant, the community, and the local planning authority.

## Will we need planning permission?

Almost certainly, yes. To get it, you'll have to show that you have carried out all of the required studies, and properly consulted with all those who might be affected by the turbine. This is no small task, but plenty of people have done exactly this.

If the consultant whom you have engaged agrees with your initial assessment, he or she will then need to do some more detailed site assessments. What these will be will vary a lot, depending on the scale of the project: in essence the bigger the project and the bigger the turbines the more studies and assessments will be needed.

First of these will be to make an accurate measure of wind speed and other factors like wind direction. This will involve erecting a mast that supports all the measuring equipment at the height that the proposed turbine will be. For small turbines in a location that obviously has a good wind resource, this step might not be necessary.

If you do put up a mast, this will be a critical moment in your project, because the mast itself will attract attention. If you haven't already gone public with your plans, people will now want to know what's going on, and this is the point at which an anti-wind campaign can take off very quickly. You must therefore have a comprehensive plan for consultation, for sharing the data and keeping people informed.

Your local authority will require several pieces of documentation. If you're hoping to install a large turbine, you may need a full Environmental Impact Assessment (EIA), but at the very least you'll need to carry out various studies including a landscape impact, and an investigation into the ecology of the site, in particular bats and bird population studies.

## How do we get a grid connection?

You'll need to work with the grid operator to ensure that your proposed connection is suitable. Small projects can often be connected to the grid locally, just by using compliant equipment, although you'll still need to agree this with the local grid operator. Medium to large wind turbines will most likely need a new line put in to the nearest connection point or substation.

If your project is in a very rural area, or is reasonably large, there may be a cost to upgrade the grid to take the new load. Dialogue with the grid provider should be started early and you should keep them up to date so that any costs can be calculated into the project finances. Upgrading the grid to accept a new connection can be quite time-consuming, and in some areas you may need to put your project in a queue for this, so early and regular communication will help minimise this period.

## How long will all this take?

Quite a long time. You'll need to measure the wind speed for at least one full year. If timed correctly, the various other studies can be done in the meantime. The bat survey, for example, would be carried out between May and September. You should reckon on a minimum of two years to complete all studies and public consultation.

You should then estimate another year as a minimum to secure planning permission, including the need to appeal if necessary. You'll also need this time to secure project finance for the full installation.

When you add this all up it's pretty daunting, but it is essentially a series of smaller, discrete tasks and costs, and you should treat the completion of each stage as a cause to celebrate. Also remember that what you learn from your process, whatever the outcome, will be useful learning for other communities trying to achieve the same, so publicise your progress, warts and all!

## How much will all this cost?

This obviously depends on the scale of the project, but the sums are not insignificant. Before you even think about buying a turbine or preparing the site, you'll need to secure funding to pay for the specialist support like consultancy and impact studies. A good wind site will repay this investment, but this is risk capital – if you fail to get planning permission, you will not recoup these costs. The Rural Community Energy Fund (RCEF) in England and the Community and Renewable Energy Scheme (CARES) in Scotland can both help with these costs.

The purchase and installation of a single large wind turbine, say a 1MW machine, along with all associated costs of consultation, planning, legal advice, grid connection, and so on, is probably in the order of £1.5m.

This sounds terrifyingly large. However, remember that, for a good wind site, the future revenue from the electricity sales is likely to be in the region of £8m over the lifetime of the turbine – around 25 years – making it an attractive investment for a bank, whom you would work with to arrange project finance. And your community would benefit from an income stream that could

Part 2

pay for other local projects and benefits like playgrounds or insulation schemes.

## Case study of small wind project – Uig Community Shop Wind Turbines

On the far west coast of the island of Lewis in the Outer Hebrides, Uig is one of the UK's most remote communities. In 2003 Uig Development Trust was formed to look into the feasibility of the community taking ownership of the local shop, which had recently been put up for sale. The trust bought the shop in 2004 using a combination of shares and grants and then made a number of improvements to it and the services it offered to the local community. By 2009 the turnover had more than doubled and with two full-time and six part-time employees, it was one of the largest employers in the area, providing crucial services for the community as well as forming a significant social hub.

**The wind turbine in Uig.**
(photo: Uig Community Shop)

The development trust then successfully applied for a grant to increase the size of the premises so there would be more shop space, more storage, a cafe and a launderette. Being aware that the energy consumption would be high, the trust wanted to make the extension as energy-efficient as possible and looked into the possibility of using a recovery system to reuse heat from the chill cabinets and the launderette in the shop. However, this proved not to be suitable so instead the trust looked into other options and, based on the site location and the electrical demands of the shop, chose to install two small wind turbines.

The only problem with this was that the local electricity network was only single-phase, while the turbines would require three phases. Making this upgrade added significantly to the cost of the project but had the benefit that higher-duty equipment could be installed in the launderette.

The turbines used for the project were Evance 5kW and the total cost was just over £88,000. This amount was sourced from a Scottish government CARES grant together with a grant from Comhairle nan Eilean Siar and the Big Lottery.

The turbines were installed and commissioned in June 2011. Between them, they generate around 28,000kWh of electricity per year and save around 12 tonnes of $CO_2e$ per year and the shop earns around £8,000 from the Feed-in Tariff. This all increases the long-term sustainability of a vital local service in a remote community.

**The wind turbine at Hockerton, Nottinghamshire.**
(photo: Hockerton Housing Project)

## Case study of medium wind project – Sustainable Hockerton

In contrast to Uig in the far north-west of Scotland, Hockerton, a small parish close to the market town of Southwell, is in the middle of England. Sustainable Hockerton[2] (or SHOCK as it's known) was set up in 2006 with the aim of making the village more sustainable and reducing the amount of carbon released into the atmosphere. The group defined a number of specific objectives and among them was the aim of meeting the village's energy needs from renewable sources. The group

also felt that a focus for their work was required, plus a source of secure revenue to support the other activities. Having a wind turbine in the village and selling the electricity generated to the grid would seem to meet all these objectives.

Before going any further with the project. the group felt it was important to consult with the rest of the community so a questionnaire was developed and distributed to all the households in the village. Most of these were returned and the vast majority indicated support for the project, with many of them also showing a willingness to invest in it. After a number of local landowners had expressed an interest in hosting the turbine, a suitable site was found. A planning application was then prepared and planning permission was granted in June 2008.

In order to achieve the aim of meeting the village's energy needs, a fairly large turbine was needed so the planning application was based on a Vestas V29 turbine, which has a rated output of 225kW. It was calculated that this should produce around 330 megawatt-hours (MWh) per year, while the domestic consumption in the village was estimated at 275MWh. These turbines, which are no longer made, have a blade diameter of 29m and a hub height of 32m so while they're not among the largest onshore turbines, they're still substantial structures.

The group were able to locate an unwanted Vestas V29 turbine at an industrial site in County Durham but found themselves in a competitive situation with a number of other potential buyers. Luckily, they were able to raise the necessary funds from local investors to secure it relatively quickly.

The full cost of the turbine and installation was estimated at around £225,000 and although they applied for a number of grants, they were unsuccessful so raised the full amount from investors. These investors were focused locally, with over a third of the capital coming from Hockerton itself and over two-thirds in Nottinghamshire, with the remainder coming from other parts of the UK.

The legal structure that was used for the project was an Industrial and Provident Society (IPS), which was christened Sustainable Hockerton Ltd. As described in Chapter 13, the advantages of being an IPS are that it is democratic and it allows members to invest in the project.

Construction on the site started late in 2009 and the turbine started exporting electricity in January 2010. The project took place just as the Feed-in Tariff was being introduced for small and medium-scale wind projects in the UK.

By 2013 the turbine had generated over 970,000kWh of electricity and had saved the equivalent of 507 tonnes of $CO_2$e. It was also paying around 7% on average each year to investors and £25,000 had been raised for other projects in the village that support sustainable development. Households in the village were offered £200 each towards "quick wins" including warmer curtains, insulation, water-saving devices and energy-efficient lighting.

Chapter 7

# HYDROPOWER

Hydroelectric systems work by transferring the energy of moving water into electricity using a turbine and a generator. They do this by exploiting the water's potential energy (the water's fall) along with some kinetic energy (the water's flow). The fall of a river is referred to as "head", and it's a term you'll hear a lot when we talk about hydro. You'll notice that we'll mention it far more than flow for important reasons we'll come on to later.

In essence, all small hydro systems take some water out of a river or stream, it then passes through and turns the turbine, and re-enters the river lower down. They don't have very high dams or reservoirs and so don't significantly alter the nature of the river in the way that large hydro schemes do.

The smallest hydro systems commercially available are about 600W. One of these would provide most of the electricity for a single, average-sized house over the course of a year. A fairly typical size for a community hydro system is 30kW. This kind of turbine could supply the electricity needs of around 50 average homes.

At the top of the scale of "small" hydro are systems of about 5MW. It's a slightly misleading term, as 5MW is actually quite big – the equivalent to two large wind turbines and enough to supply the electricity demand of

around 8,000 homes. However, the term serves to differentiate them from the industrial-scale hydropower schemes of 1,000MW or more that can involve the damming of rivers and flooding of entire river valleys.

Unlike wind power and solar power, hydropower generates energy almost all the time, varying slightly with seasonal alterations in flow. This means that, over a year, a 1MW hydro installation would generate far more electricity than a 1MW wind turbine or solar array because it would be generating more of the time, at or near to its maximum capacity.

Hydro is also a fairly low-maintenance technology and has a longer working life than other renewable technologies. Typically they'll last 30–50 years, as opposed to 25–30 years for wind and solar. As such, although the upfront costs can be high, income once up and running is steady, predictable and long-lasting.

## How a typical small-hydro system works

Some of the water is diverted from the river into a specially dug open channel, while the rest continues downstream. All hydro systems need to be designed to

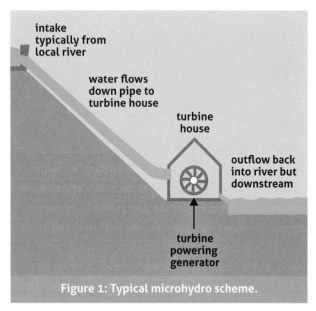

**Figure 1: Typical microhydro scheme.**

Labels in figure:
intake typically from local river
water flows down pipe to turbine house
turbine house
outflow back into river but downstream
turbine powering generator

make sure that the part of the river that doesn't pass through the turbine always has adequate flow for fish and other aquatic life.

The diverted water then passes through a tank where silt and small stones are caught to prevent them entering the machinery. If the river is relatively silt-free, this may not be necessary. There's also a grid structure here that sieves out floating debris, like branches and rubbish. Sometimes the water flows directly to the turbine itself, otherwise it first enters a section of pipe, which can be quite long if the turbine house is some distance from the intake.

Turbines are large pieces of machinery – a 1MW turbine would weigh several tonnes. The water turns the turbine as it passes through or over it, depending on the design. They come in many forms, such as Archimedes screw, Pelton wheel and crossflow. Different designs suit

different circumstances. Archimedes screws, for example, are more suitable in rivers with a low head.

Next to the turbine is a generator. As the turbine turns, it powers the generator, which generates the electricity. The system will also need an electricity substation that will feed the energy generated by the hydro plant into the grid. After the water's passed through the turbine house it re-enters the river.

## Where can we put one?

Not all rivers are suitable for hydro. And as I've said, this is because they work by exploiting the water's potential energy - the water's fall or head - along with a smaller amount of its kinetic energy, the water's flow. The fact that it's about head rather than flow is critical. Many people naturally assume that any fast-flowing river would be a good site for hydro power. However, in the absence of significant head, there's unlikely to be sufficient energy to exploit, even in a fast-flowing river. The head of the river can be natural, for example a stream flowing downhill, or it can be artificially generated by building a weir or dam.

What this adds up to is:

- Large rivers with significant falls are the best sites for hydro because they have high head and high flow.

- Rivers with low head and low flow are no use at all.

- Rivers with high head but low flow can be a decent bet, but only as long as they do actually flow most of the time. Some small, fast-flowing rivers dry up during the summer.

- Large rivers like the Trent or Severn, which flow over flattish terrain, are characterised by low head but very high flow. These can be good sites, but you have to extract quite a lot of water to achieve decent generation, so might struggle with costs, planning and licensing.

If you're prospecting for a site for a hydro plant, a good question to ask is if there is, or ever has been, a watermill on your river. If so, then it's definitely worth an initial investigation. The river Frome in Somerset once powered a string of mills that were used in the wool trade with some of them dating from Saxon times. Today, some have been brought back into life as hydro plants.

## What about fish and other river life?

One of the main questions people ask about hydro is "What happens to the fish?" And it's a good point; fish passing through turbines are sometimes killed. So most hydro plants are designed to keep fish out, using bypassing channels, sieve-like screens, eel-pipes, and even "fish ladders". These are a series of small pools, one above another, that migrating fish like salmon can use to jump up and bypass the turbine.

All hydro schemes, no matter how small, must be licensed by the relevant government environmental agency that will expect to see plans on how the river's wildlife will be protected, both when the scheme is being built and after operation. And this is not just fish we need to worry about, but amphibians, crustaceans, mammals, insects and plants too – the whole river community in fact.

## Hydro, FITs, CFD and ROCs

Like wind, the position with regard to financial support for hydro projects could be a bit confusing, with the FIT, CFD and ROC all being applicable, depending on the size of the project and the date it will be commissioned. However, in reality it's relatively straightforward, with FITs being the best mechanism for any project below 2MW in installed capacity, which is going to account for the vast majority, if not all, of community hydro projects (at the time of writing, the largest community hydro project in the UK is in Callander in Scotland, with an installed capacity of just over 400kW[4]).

Projects between 2MW and 5MW have the choice of either FIT or CFD, while projects above 5MW have only the CFD to worry about, unless they're commissioned by March 2017, which is unlikely for any project that's not already well under way.

More information about how FITs work in practice together with a bit more information about the CFD and ROC can be found in Chapter 12.

## A hydropower project

So, you and your group wish to investigate the possibility of a small hydro project. What next?

You will almost certainly, at some point, need to employ a consultant. Hydro systems are very complex to design and it's very unlikely, even with significant engineering and project management expertise in your group, that you'll be able to do this alone. However, even before you get to this stage, there are some useful investigations you could undertake.

## Head and flow

First and foremost you need to identify a suitable site; one that would make the project economically viable. Above all, this means one with the right head and flow. If these factors aren't right, it's back to the drawing board.

So what are head and flow? The head is basically a measure of how far the water drops compared with how far it flows. If your river runs over pretty steep terrain, you can make an initial estimate of head from the contour lines of the largest scale of Ordnance Survey maps. For rivers with lower head, running on flatter ground, it's more complicated and you'll need the services of a qualified surveyor with a theodolite.

Understanding the flow pattern of your river is also important. What you're looking for is a river that flows fairly consistently over the year. What you don't want are sometimes described as "flashy" rivers – those that have a reasonable average annual flow, but characterised by "flash floods" from periods of high rainfall.

And look out for seasonal variations. You won't find it economical to put a hydroelectric system into a river where summer flow is reduced to almost nothing.

You'll need data going back over as many years as possible to identify the flow patterns and plot these on a graph called a "Flow Duration Curve". This is one of the main tools for instantly assessing whether a site will be any good for hydro.

Fortunately, there is a low-cost way of accessing some of this data. The UK National River Flow Archive[1] at Wallingford in Oxfordshire has records of river flow data gathered from more than 1,300 monitoring stations in the UK. The staff at the archive are able to tell you how near your site is to one of their gauging stations, and for a fee of between £50 and £300 they can generate the flow data that you need, including your Flow Duration Curves.

You can use these to make an educated guess at whether your site is worth discussing with a consultant. Note that if your site happens to be some distance from the nearest monitoring station, the archive data may not be very accurate, in which case there are consultancies with software that can model the difference for your site, at a cost of a few hundred pounds.

Whatever the outcome of your investigations into flow data, further measuring – possibly for a year or more – may be necessary for sufficient accuracy.

## What other site characteristics can you assess on your own?

Once you've identified a suitable site, you need to find out more about it.

1. Who owns the land? And do you think you could enter into a purchase or lease with them?

2. **Is there a reasonable connection to the grid?**

   If you are miles from any grid access or properties that might need to use the electricity, generating hydroelectricity is probably not going to be economical.

3. **Is the site accessible to machinery?**

   And we're talking big lorries and earth-movers here. A smallish turbine can weigh several tonnes.

4. **Who may be impacted by the project?**

   The river-using community probably includes anglers, kayak clubs, wildlife enthusiasts and more. Start building relationships.

5. **Is there, or has there ever been, a mill on the river?**

   If the mill house is still there, it could potentially be reused to house a turbine, and this may make gaining planning permission easier.

   You need to make sure that the run of the river hasn't been significantly altered since the mill was in operation, for example by dams or weirs being built elsewhere.

   Speak to people who have lived locally for a long time, and consult old maps and photographs.

## Will we need planning permission and other licences?

The short answer is yes, to both. Planning permission will be required for any works along the river – building the weir and the turbine house for example. Even if there's an existing mill house, it's very likely that you'll need permission for the required alterations, and there is a reasonable chance that an existing mill house may be listed anyway. If it is, you'll also need Listed Buildings consent.

Because all hydro projects will require works near riverbanks, and altering the flow of the river, you will almost certainly be asked for an environmental assessment to accompany your planning application, so you'll need the services of an ecologist.

### Environment Agency or SEPA licences

Unlike other renewable technologies, where most of your permissions will be granted by the local planning authority, hydropower has another layer of complexity. You'll also need several licences from the Environment Agency[2] (EA) or Scottish Environment Protection Agency[3] (SEPA) in Scotland. These include:

- An abstraction licence.

- An impoundment licence.

- Flood defence consent.

- Freshwater fisheries approval.

They are not expensive; however, each licence requires supporting evidence, such as migratory fish surveys. Such studies can only be done at certain times of the year, which you must build into your timetable. Many hydro projects fail or are delayed because groups badly underestimate how long it will take to get these licences.

And please, do not view the Environment Agency or SEPA as an obstacle to be overcome. They have a statutory duty to protect and enhance the UK's rivers, and to manage flood risk. They're not there to try to stop your project, but to protect the national river network. Their remit and responsibility is huge, and your project is a tiny part of it. Be prepared to work with them, and to their timescales. Talk to them early and often, and you will get what you need in the end.

## Grid connection

You will also need to work with the grid operator to ensure that your proposed connection is suitable. Small projects can often be connected to the grid without permission, just by using compliant equipment. If your project is in a very rural area, or is reasonably large, there may be a cost to upgrade the grid to take the new load. Dialogue with the grid provider should be started early and you should keep them up to date so that any costs can be calculated into the project finances. Upgrading the grid to accept a new connection can be quite time-consuming, and in some areas you may need to put your project in a queue for this, so early and regular communication will help minimise this period.

## How long will all this take?

Hydro is said to be the most time-consuming of all technologies to get from inception to installation, so be realistic with yourselves and with others. For even the smallest of systems you should plan for a minimum of 3 years. Don't be put off – the rewards can be huge. If your site is a good one, hydro is a very reliable, predictable and long-term energy generation option; operating lifespans of 50 years are not uncommon.

## How much will all this cost?

A small system in a straightforward site could cost as little as £10,000 to get to the point of installation, with perhaps a further £50,000 to install. Bigger community systems could be in the low millions.

Remember, though, part of the work you'll do with your consultant will be to establish the economic feasibility of the project as a whole. Although a project cost of more than a million may seem out of reach of a community group, this will be offset against significant operating income from electricity sales. And that should form an attractive investment for a bank.

# Hydro case studies

To illustrate some of the different aspects of hydro projects, I've used two different examples as case studies. The first at Whalley in Lancashire is an example of an Archimedes screw funded by a local share issue, while the second in Talybont is an example of a Pelton wheel turbine. As well as both being hydro projects, the common element they share is local people spotting that there was something already in their local area that could be exploited for the good of the community and the wider world.

## Whalley Community Hydro[5]

Whalley is a village on the banks of the River Calder, not far from Clitheroe in Lancashire. The community hydro project is a 100kW Archimedes screw that went live and starting exporting electricity to the national grid in November 2014. It's expected that the scheme will generate around 345MWh each year for the next 20 years.

The community hydro project grew out of Transition Town Clitheroe[6], a local Transition initiative. It was one of a number of different types of projects the organisation were looking into at the time. There's a man-made weir across the river in Whalley, which guides water into a channel that historically fed a water wheel for the corn mill. Some of the members of Transition Town Clitheroe realised that this weir could mean that Whalley was a suitable site for a small hydro scheme.

The river has a large catchment area and what's described as a "high base flow", meaning that the flow is relatively consistent and not "flashy". This high flow is combined with a relatively low head so it's well suited to an Archimedes screw, which has the additional benefits of being efficient, robust and fish-friendly.

The team that developed the hydro came together in February 2010 in what's best described as an organic process – people offering their expertise when they heard about the project. Soon after coming together the group attended a workshop in Manchester run by H2OPE, a social enterprise that specialised in helping communities create hydro schemes (H2OPE sadly no longer exists). Armed with information from this workshop, the team started putting together plans for the scheme. By August 2011 ecological surveys had been completed and the team had formed an Industrial and Provident Society (IPS) as the legal vehicle for the project, plus they had reached initial agreements with the necessary land-owners.

**The Archimedes screw being installed at Whalley in Lancashire.** (photo: Chirs Gathercole)

By February 2012 planning permission for the scheme, including approval from the Environment Agency, had been received. In May 2012 construction work in the form of stabilising the weir started and by September 2013 approval had been received from the Environment Agency for water abstraction and the fish pass.

Although the project received a number of small grants, it was clear that grant-funding for the capital cost of the project – paying for the screw itself, the turbine and all the civil engineering work and so on, which was going to amount to around £750,000 – was not going to be possible. This was why they formed an IPS so that they could raise the money through local people investing in a share offer.

In February 2014 a first tranche of funding was raised from a share issue and in May of that year contractors on site began the main part of the construction work and the scheme began generating in November 2014.

Part 2

At the time of writing, around £520,000 of the total cost of £750,000 has been raised from the share offer (with the remainder being paid for from loans). However, the aim is to continue with the share offer until all £750,000 has been raised and all the loans have been repaid.

The distribution of the investment has been quite highly localised to the scheme. Of the money raised via the share offer, 50% has been from people in the Ribble Valley so very local to the project. A further 25% is from wider east Lancashire, 15% from the north-west of England and the remaining 10% from other parts of the UK.

As well as generating green electricity, and a return for the investors, the plans are for the scheme to have wider community benefit. Any operating surpluses will be used for schemes to reduce carbon emissions, assist in adaptation to low carbon living and increase energy security in Whalley and the surrounding area. Funds may be given as grants or loans to organisations such as village halls, schools and sports clubs as well as to residents and businesses.

## Talybont-on-Usk Energy[7]

In contrast to Whalley, which has only recently gone live, the hydro project at Talybont in the Brecon Beacons is the first community hydro scheme in Wales, with the turbine installed in 2005 and formally launched and supplying electricity to the national grid in 2006.

This hydro scheme consists of a 36kW Pelton wheel turbine working off the "compensation flow" from Talybont Reservoir, which was built in the 1930s to provide drinking water for the Newport area. Welsh

Water, which operates the reservoir, is required to keep the River Caerfanell at statutory levels. To do this, water is taken from the reservoir and fed into the river; some of this "compensation flow" is channelled through the turbine. When the reservoir was first built, a turbine operated on the site to provide electricity for the water works but when the valley was connected to the national grid, the turbine was decommissioned.

In contrast to the scheme at Whalley, the Talybont turbine operates with a relatively low flow but with a much higher head (in the case of Talybont, it's about 24 metres). This makes it ideal for a Pelton-wheel-style turbine rather than an Archimedes screw.

A group in Talybont was formed as long ago as 2001 when the Brecon Beacons National Park Authority facilitated public meetings in the park, looking at the potential for communities to develop renewable energy projects. A subsequent feasibility study paid for by the park authority's sustainable development fund identified installing a new turbine in the existing turbine house as a potential project.

Over the next two years, the group set up Talybont-on-Usk Energy Ltd, a company limited by guarantee as the legal vehicle for the project, negotiated the lease with Welsh Water, which owned the turbine house, and commissioned a detailed design and implementation study for the new turbine. This work suggested that the total cost of the project would be around £92,000.

The group secured grant funding for the project from the national park, the UK government's Clear Skies fund and an ERDF Objective 2 fund, so that by January 2005 they were able to order the new turbine and by Novem-

**The generator at Talybont, in the Brecon Beacons, Wales.**
(photo: Talybont Energy)

ber of that year the turbine was fully installed and commissioned.

The scheme produces between 220 and 250MWh annually. This in turn saves around 120 tonnes of $CO_2e$ and generates a gross income of £30,000 for Talybont-on-Usk Energy to invest in other sustainable living projects in the local community. In total to date, the turbine has generated over 1,000MWh.

One of the most impressive things about the Talybont scheme is what has been done with the income generated by the turbine. Like a number of other generation schemes around the country, the income is used by the group to run a wide range of other energy-related projects including:

- An electric bike trial, which resulted in six households buying their own electric bike.

- A Key Stage 2 educational package that links directly into primary school National Curriculum subjects like Geography, Science and Citizenship.

- The installation of 4kW of photovoltaic panels on the roof of the village hall in 2011 that have themselves generated over 10MWh of electricity to date.

- The creation of a community fund for individual projects, including the creation of a local small-scale biodiesel plant from locally sourced waste cooking oil.

- The installation of a further 4kW of photovoltaic panels on the village hall together with air-source heating.

- A community car-share project that has been operating since 2011 and runs two cars – one electric (charged from the photovoltaic panels at the village hall) the other powered by biodiesel (created at the local biodiesel plant).

Part 2

Chapter 8

# HEAT PUMPS

This chapter is all about helping you understand how heat pumps work, and what different types of heat pump there are, so that you can make an informed decision about what could be right for your community.

A common misconception is that heat pumps use geothermal energy – hot springs or hot rocks in the earth. They don't. What heat pumps actually do is to capture solar energy – energy from the sun that has slightly warmed the earth, a body of water or the air.

As such, you will hear people mention ground-source heat pumps, water-source heat pumps and air-source heat pumps.

## How do they work?

Although it might seem hard to believe, even on a cold winter's day, the temperature a metre underground or at the bottom of a river remains surprisingly constant, at about 10°C. Heat pumps are able to extract this warmth for you to make use of.

### Ground-source and water-source heat pumps

Ground-source and water-source heat pumps do this through a series of coils or pipes. A liquid which is slightly below 10°C is pumped through the coils and picks up heat from the surrounding earth or water. It is then passed through a device called a compressor, which pressurises the liquid and, by so doing, raises its temperature to about 40°C or a little higher – a little warmer than a baby's bath. This heat is transferred to the water in your heating system via a heat exchanger. The liquid, now cold again, goes back to the earth or water to extract more heat, and the cycle begins again.

The compressor and heat exchanger are housed in a unit about the size of a large fridge. Electricity is needed to power the compressor, so heat pumps, which are powered from mains electricity, are not completely carbon-neutral. However, in the right place, ground- and water-source heat pumps can turn every kilowatt hour of electricity into 3 or 4 kWh of heat energy so they are a very good way to reduce carbon emissions from heating. And if you source the electricity needed from a renewable source such as a wind turbine or solar panel then so much the better.

For ground-source heat pumps, the main constraint is space; do you have room to dig trenches for the coils? If not, the coils can be installed vertically – but this is considerably more expensive, as it requires drilling deep

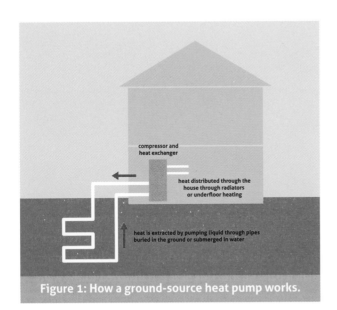

**Figure 1: How a ground-source heat pump works.**

Labels within figure:
compressor and heat exchanger

heat distributed through the house through radiators or underfloor heating

heat is extracted by pumping liquid through pipes buried in the ground or submerged in water

boreholes. A rocky site obviously brings its own problems.

Water-source heat pumps work where there is a reasonably high flow of water; this keeps the temperature more even. Some larger systems need a licence from the Environment Agency or SEPA because they extract water from the river or canal.

### Air-source heat pumps

Air-source heat pumps work rather differently. Mounted directly on an external wall of a house, they work – and look – rather like air-conditioning units, only they run in reverse. They don't use coils, which makes them easier and cheaper to install, but they're usually less efficient, producing perhaps only 2.5kWh of heat for every 1kWh of electricity. Because they rely on the outdoor air temperature, air-source heat pumps have to work harder when it's cold, so use more electricity during the winter months. For this reason, although you'll find them in buildings throughout the UK, the best efficiencies are achieved in areas with milder winters.

## Why don't we all have one, then?

If heat pumps are so efficient, why don't we all have one? Well, if your building is connected to mains gas, the economics don't really stack up. This is because mains gas is about one-third of the price of electricity, so there's no point in substituting a gas central-heating system with a heat pump – unless your aim is to save carbon at all costs. Sure, you'd be using one unit of electricity where you previously might have used three units of gas, but it would cost you the same. For this reason heat pumps tend to be used in buildings that are off the gas network and where the building would otherwise use more expensive heating fuels, such as oil or electricity.

As you can see, the successful use of a heat-pump system depends on a variety of factors, not least the efficiency of the building itself, and what other source of fuel you have available.

## A heat-pump project

So, you've decided to explore further the possibility of setting up a heat-pump project for your community. What are your next steps?

First of all you should draw up a heat-demand profile of the building or buildings that you intend to install the heat pump in. In other words, how much heat does the

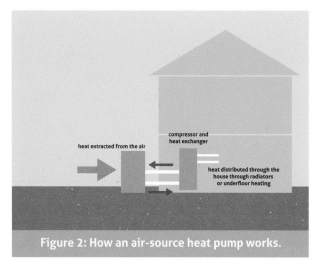

**Figure 2: How an air-source heat pump works.**

heat extracted from the air

compressor and heat exchanger

heat distributed through the house through radiators or underfloor heating

## Sizing the system

It's important that your heat-demand profile is as accurate as possible because it's this that decides the size of the system that is eventually installed. If you end up with an oversized heat pump – one that can provide more heat than the building requires – it will simply cost you more for no additional benefit.

An undersized heat pump won't heat the building adequately in the colder times of year and will have to rely on a back-up system, such as gas or electric heating. In some cases it might actually be more economical to undersize a heat-pump system and rely on some sort of backup, but you need to be sure that this is done by design and not by accident!

## Powering the heat pump

Heat pumps use electricity, and there are various options for buying this. You should discuss this with the heat-pump contractor at an early stage.

If your aim is to be as near to zero-carbon as possible, you could install solar panels or a small wind turbine to compensate for the electricity that the heat pump uses. This will of course increase the cost of the installation but could mean that you benefit from Feed-in Tariff payments as well as the Renewable Heat Incentive.

Remember, though, that since the sun may not be shining or the wind blowing when you want the heat pump to work, you'll still end up using mains electricity. In which case you may want to consider "green" electricity certified under the Green Energy Supply Certification Scheme.[2]

building use and when is it required – what hours of day and in which months of the year? The data that you collect will be critical. Without it, you won't be able to specify a heat pump that is fit for purpose.

Your energy bills will tell you how much heat you use. However, you may have to rely on your own knowledge of the building and how it is used to work out at what times of the day – and for how long – the building is heated.

Ideally, your heat-demand profile will cover a full year, taking seasonal fluctuations into account.

You should also consider the impact of any proposed changes to the building – for example extensions that might be planned, or significant changes to patterns of use.

And all this is assuming that the building is – or will be – well insulated; you should think twice about installing a heat pump in a poorly insulated building.

A further option is switching to an off-peak tariff such as Economy 7. This will save you money if the pattern of use is suitable – for example if the heat pump is mostly on in the evenings.

## Contractors

When it comes to engaging a contractor, as with any other major building project, you should get at least three quotes. You should check out references and may want to refer to the website of the Heat Pump Association.

In particular, if your system involves drilling a borehole instead of digging a trench, make sure that the risk of unforeseen technical difficulties lies with the contractor and not your group. Ensure that this is clearly stipulated in the contract.

## Will we need planning permission?

Ground-source and water-source heat pumps raise few planning issues, as so much of the system is hidden from view – buried in the earth or submerged in water. In most cases they are permitted developments. They're so inconspicuous in fact that you may want to raise awareness of the technology in your community build-ing through the use of display boards or posters. Air-source heat pumps are more conspicuous, but are increasingly accepted by planning departments.

However, you will still need to consult your council for Listed Building Consent where relevant, and ensure that all heat pumps comply with Building Regulations.

**A drilling rig in action.**
(photo: Katherine Cowtan)

There are noise issues with air-source heat pumps, which will affect where they can be located so as not to disturb neighbours. You should talk to your system designer and your neighbours about this at the earliest opportunity.

## Heat pumps and earning money from the Renewable Heat Incentive

As a technology that produces renewable heat, heat pumps are eligible for the UK government's Renewable Heat Incentive (RHI).

Part 2

As described in Chapter 12, the RHI is a mechanism introduced by the government to encourage the installation of heat-generating renewable technologies. Both ground source and air source heat pump installations are included in both the Domestic and Non-Domestic RHI.

## Heat pump case study – Waternish Village Hall

Because they require very little maintenance and are relatively easy to install (air source in particular), heat pumps can be a good source of heat for community buildings.

Waternish is a peninsula situated on the north-west of Skye, overlooking The Minch and the Outer Hebrides to the west and Loch Snizort to the Trotternish Ridge to the east.

Waternish Village Hall was built in 1956. Being on the west cost of Skye, the weather is a constant challenge and in the intervening years, refurbishments have focused on ensuring the hall survived rather than making significant improvements. The single-storey building has a high-pitched ceiling and was heated by five electric radiant wall heaters and had very little insulation (in some places no insulation) so that the heat that was produced by the wall heaters was quickly lost. Over the years, the committee who run the hall found the heating system both expensive and inefficient, and were beginning to consider the options available to them.

An energy appraisal funded by the Scottish government's CARES programme (see Chapter 14) looked at the various options that were available, given the age and construction of the building and its situation. This appraisal looked at both energy-efficiency measures that could be taken as well as renewable energy options.

In making the decision about what route to take the committee also had to take into account that there is no caretaker for the hall, so whatever solution was chosen needed to require minimal maintenance. Another factor complicating the decision was that improving the heating system was only part of a wider renovation project for the hall, which involved a number of different funders so funding deadlines also needed to be taken into account.

Ultimately, the committee decided to improve the energy efficiency of the building by installing cavity-wall insulation and replace the existing wall heaters with an air-to-air heat pump. The committee considered other renewable options like a wind turbine and a small biomass system but their key aim was to end up with a hall that was easier to heat so the simplicity of the air-source heat pump was preferred.

The device installed was an air-to-air heat pump and the total cost for the project, including the insulation, was just under £11,000. Half of this cost was met by a Scottish government CARES grant, with the remainder coming from LEADER* funding and the hall's own funds.

Although, compared with some of the other case studies in this book, this example of community renewables is very simple and straightforward, the benefits are considerable, with the community having a warm, comfortable facility, which also has affordable heating costs.

*An EU programme providing funding for organisations to begin or expand their operations in rural areas.

## Chapter 9

# BIOMASS

In an energy context, "biomass" refers to burnable material derived from wood or other plants, and usually one of the following:

- Wood, either in the form of logs, woodchips or wood pellets.

- "Woody" materials such as specially harvested fast-growing crops like willow or poplar.

- Fast-growing tall grasses such as elephant grass.

- And finally other by-products such as sawdust or wood chippings from sawmills or wheat straw from farms.

These fuels are considered to be sustainable, as the amount of carbon dioxide emitted when they're burned is recaptured by the fresh growth of new crops or trees.

Biomass fuels are supplied in a number of different forms, from logs, wood pellets and woodchips, to sawdust or briquettes. These all vary in size, weight and other physical properties and so are used in a range of different equipment.

## How is energy generated from biomass?

There are several ways to generate energy from biomass. Almost always this involves combustion – or, put simply, burning – and the majority of community-sized biomass systems use this process. The heat produced by burning the woodchips, logs or other biomass is used directly for room or water heating, or in large-scale systems to produce steam, which is then converted into electricity using a steam turbine. Biomass can also be converted into liquid biofuels for transportation – but this isn't relevant to us here.

In the UK, power stations burn large volumes of biomass to generate electricity.* At the other end of the scale biomass is also used to provide heat in domestic housing using open fires, wood stoves or boilers supplied by wood pellets. In between are the medium-scale biomass systems, such as those used to heat community halls or

---

\* Burning biomass at this scale is controversial. For example, the Drax power station is being converted so that three of its six units will burn biomass rather than coal. It's expected that the vast majority of the fuel will be imported from the US and Canada and a special facility is being built at Hull for this. Many people believe that burning biomass at this scale and importing the fuel in this manner is a long way from being sustainable.

small businesses, or to heat several different buildings that share a district heating system. Systems known as Combined Heat and Power (CHP) use biomass to generate electricity via a steam turbine or other similar devices, but they also use the waste heat from this process to provide heating. This option can be the most efficient use of biomass resources.

## Types of biomass system

Biomass systems of the size found in homes or used as part of a community energy project are nearly always simply used for heat generation.

### Log stoves and boilers

The simplest use of biomass fuel is found in the humble log stove. Considerably more fuel-efficient than open fires, these come in various shapes and sizes, and if you're keen to heat your house with wood fuel on a limited budget, this is probably your best bet.

**A typical log stove.**
(photo: iStockphoto/joshuaraineyphotography)

Log-fired boilers are a logical step up from stoves. They range from systems designed for hot-air space heating – found in some workshops and similar premises – to boilers designed to run domestic heating and hot water.

### Pellet boilers

Pellet boilers do something similar, only instead of using

**Wood pellets.**
(photo: iStockphoto/ yotananchankheaw)

logs, they burn manufactured pellets. These are smooth and dry and look like pony nuts. They are usually fed into the stove automatically from a hopper. Again, they are becoming increasingly popular in homes. Pellet boilers come in all sizes, from small domestic appliances to a big power station. Many domestic pellet boilers have an internal hopper, or are filled manually from bags. The bigger utility boilers, say for a school or leisure centre, have separate, bulk storage hoppers, which can often be filled just once a year via a long pipe.

### Woodchip boilers

These are similar to pellet boilers, but tend to be larger because woodchips require more robust feed and burning systems. You'll see them in settings such as blocks of flats, visitor centres, offices and hospitals. Woodchip quality is vital – boiler manufacturers will specify a chip size and moisture content, and you should make sure that your contract for fuel supply specifies this.

Woodchips can be produced from round wood by using specialised machinery, which produces a uniform size of

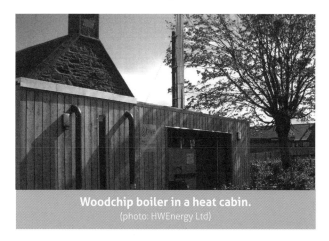

**Woodchip boiler in a heat cabin.**
(photo: HWEnergy Ltd)

chip. Since woodchips require less processing than pellets and less manual handling than logs, they can be an extremely energy-efficient use of biomass. Country estates, farms or other sites that have access to woodland can supply their woodchip boilers themselves, and either do their own chipping or use a specialist contract chipper. A number of buildings can also be connected to the same biomass boiler via a district heating system.

## Storage and supply

As you'll have realised by now, if you want a biomass system, you'll need somewhere to keep the fuel. Unlike other types of renewable energy, it doesn't come down a cable or through a pipe! Sometimes this will mean taking over some or all of an existing building or putting up a new one where logs, pellets or woodchips can be stored.

For woodchip, the storage area needs to be suitably airy, so the material can dry out naturally. Freshly chipped wood has a high moisture content, and if it is too wet it will start to rot and your system will burn much less efficiently.

Where possible, try to minimise the distance your fuel has to travel to get to you. This will reduce the cost of transport and the associated carbon emissions. Large users, such as power stations, often source their fuel from abroad, where, sadly, ancient forest is cleared to make way for fast-growing biofuels. You should investigate in some detail whether your wood fuel supply is really sustainable. You might consider negotiating wood fuel or pellet supply contracts with local forestry estates or sawmills. This will also help support British rural industry.

## What about emissions?

Even if you live in a smoke-free zone, you will still be able to install a biomass boiler, as long as the model you wish to use is on the government's list of "exempt appliances".[1] However, you should still be aware of the potential of biomass systems to emit smoke along with other chemicals.

Part 2

One way to keep emissions to a minimum is to maintain the boiler in a good condition, and ensure that your fuel is not too wet. You should also ensure that the fuel is "clean" and avoid using contaminated fuel, such as wood waste, which may have been treated with chemicals.

There's also the problem of ash. The ash that falls through the grate during combustion is known as bottom ash, which can actually form a valuable fertiliser for the garden or allotment.

Individual biomass systems can range from simple pellet or log stoves through to boilers big enough to heat a community centre or school.

## Setting up a community-owned biomass boiler

Buying a biomass boiler is much like buying any other heating system – you'll want to obtain several quotes, and make sure that the installer is reputable. Check reviews, ask for references, and determine whether they are registered with the government-backed Microgeneration Certification Scheme.[2]

Even at a small scale, the first thing you'll need to consider is fuel supply – where will it come from, is it a reliable supply, and do you have the space to store it? Wood pellets require very little storage space, but are more likely to come from further afield and may not be economical in small quantities. Logs need more storage space, but you may be able to get them very locally and in small quantities.

Once you start looking at bigger boilers, the sort that can heat a school, place of worship, or a community centre, planning the installation will feel more like a project and less like the purchase of a domestic appliance. However, the first thing you need to address will still be fuel supply.

### Fuel-supply options

The Forestry Commission's Biomass Energy Centre website[3] contains a list of fuel suppliers by region and is a good first point of contact. If your building is near a farm, sawmill or estate, it's also worth contacting them to explore the options for local supply. If they don't already produce woodchip, they may be willing to if you can guarantee them a viable market. However, you'll need to make it clear that you need woodchip that meets a certain specification in terms of size, uniformity and moisture content.

### How much fuel will we need?

Starting with the heat demand of your site, it's relatively easy to roughly estimate what annual quantity of fuel you might need. Start by looking at the site's space and water heating bills over the last two years – this will indicate the typical amount of energy that is needed over a year and you can convert this into tonnages of wood fuel.

Woodchip is normally supplied with a moisture content of about 35%. One tonne of this will provide around 2,900kWh of heat energy. Pellets, on the other hand, are dryer and denser, so a tonne gives you around 4,800kWh. So if your bills show that the space and water heating of your building was around 17,400kWh per year, then you need about 6 tonnes of woodchip, or just over 3.5 tonnes of pellets.

## Delivery and storage

Woodchip and logs are likely to be delivered by lorry, pellets by van, so if you have difficult access you may be limited to pellet systems. In addition, woodchip and logs take up quite a lot of space - about 4m³ per tonne compared with just 1.5m³ a tonne for pellets. So if storage space is an issue, pellets may be the best option again. Either way, you may have to consider building from scratch or adapting an existing building for a fuel store.

## Arranging a fuel supply contract

It's important to make sure your annual supply of fuel will be of a consistent and reliable quality, free from contaminants like soil or even nails. Most important is to make sure that the moisture content and particle size of the wood is matched to the boiler specification. These normally refer to either UK CEN or Austrian Önorm standards. You should ensure that the contract for fuel supply that you establish specifically notes the size and moisture content your boiler will accept – and it should refer to those standards for clarity.

It's critical that you establish a formal contract of supply, rather than an ad-hoc approach to purchasing. This will ensure continuity of supply, fixed costs over a certain period and will redress against a supplier who provides substandard fuel. It's also worth verifying where the

wood is sourced and whether this may change in the future.

## What about the actual boiler?

The type of system you install will depend on many variables: the available fuel supply, size of the heat demand, resources for maintenance, and whether you want hot water as well as heating. All of this can be discussed with suppliers – just remember it's best to get at least three quotes to compare systems and costs.

## Should we include a heat store?

Many biomass boiler systems, especially where there's no other source of heating, come with an "accumulator" hot water tank, which acts as a reservoir, storing heat when demand is low and releasing it later. This can give you more flexibility in sizing the boiler and could reduce costs, depending on the pattern of heat demand in your building. The hot water tank can be quite large, though, so you need to discuss where it will go with the installer.

## What design of fuel store and boiler house will we need?

This depends on the delivery options available. If access is easy and there's lots of space, a simple "tipping" trailer vehicle can drop woodchips straight into an underground store. However, if you have limited access, there are ways to use an "air blower" system, where woodchip or pellets are blown in through a flexible pipe. Whatever you opt for, the fuel store will have to be secure and dry.

If you are replacing an old coal or oil boiler, it's very likely you can adapt the site for a wood fuel store – if not, you may have to build a new structure. If the site or building is listed, you might need to spend extra on hiding or designing fuel stores and boiler houses so that they're in keeping with the building. The height and location of the flue will mostly be determined by Building Regulations, so for sensitive sites your options may be limited.

## Will we need planning permission?

Unless it's a small domestic system with no significant fuel storage or flue, the answer is almost certainly yes. For systems designed for a single building, gaining planning permission should be fairly straightforward, but be prepared to be flexible. For example you may have to consider an underground fuel store if permission cannot be granted for an above-ground one.

## Do we need a backup fuel system?

Depending on the pattern of heat demand, a biomass plant can operate in tandem with a fossil fuel plant such as a gas boiler. This way the biomass boiler functions as the lead boiler supplying the larger proportion of annual heat demand, and the gas plant supplies additional heat during periods of high demand. It may be economically more sensible to do this, rather than have a bigger biomass boiler that is working under capacity most of the time. It's also fairly common for biomass boilers that are part of a district heating scheme to have backup oil or gas boilers in case the biomass system breaks down. If you're supplying heat to people's homes, this can give additional peace of mind to everyone involved!

## What are the operation and maintenance requirements?

Compared with gas, biomass systems require more work: supervising fuel deliveries, and performing regular maintenance checks for example. These are tasks for trained on-site personnel, while servicing of the boiler and associated pipework is normally carried out under contract with the equipment supplier.

Ash disposal is another issue. The task can be partly automated, but this will increase your capital costs. Most biomass boilers need their ash cans emptied weekly during peak heating periods. The amount of ash is less than 1% of the delivered biomass by weight, which means that on-site disposal is possible in some cases. For pellets, the amount is even less

## More about costs

The overall financial viability of your scheme will depend on many factors, including fuel costs, operational costs and the Renewable Heat Incentive (RHI).

A wood-burning stove for heating an individual room will normally cost up to £3,000. A biomass boiler for a large house would be up to £20,000, and a 250kW boiler for a large public building up to £200,000, including the costs of building a fuel store. You'll need to speak to installers to get an accurate figure, as costs are very site-specific.

The price of the fuel can also be variable. Woodchip is a very localised market but is currently £100 to £125 a tonne in most areas. For the convenience of pellets you pay extra – around £180 to £230 per tonne, but it's better fuel so you will need fewer tonnes. You'll need to order a minimum amount to get these prices though, normally about 3 tonnes for woodchip, sometimes less for pellets. Small quantities will cost you more than buying in bulk.

Compared with gas or oil boilers, biomass boilers use more electricity for their motors, fans, augers and pumps. The difference, particularly on large systems, can be significant, so make sure you ask your installer if their calculations have included such costs.

## Biomass and earning money from the RHI

As a technology that produces renewable heat, biomass heating systems are eligible for the UK government's RHI scheme.

The RHI is a mechanism introduced by the government to encourage the installation of heat-generating renewable technologies, and many biomass installations are included in both the Domestic and Non-Domestic RHI. Chapter 12 describes in detail how the RHI works in practice.

# Biomass case study

Although slightly more complex because it involves three different buildings (and therefore three different organisations) the heating system at St Bride's in Douglas, Scotland is in many ways typical of biomass installations in community buildings.

Douglas is a village of about 1,600 people in South Lanarkshire, Scotland. St Bride's in the village had

originally been built as a school in the late 19th century. It had subsequently served as the local church hall until it went through a major refurbishment in 1998 and became the local community centre. By 2009, it was clear that further refurbishment would be needed and a major project was undertaken to reconfigure and redevelop the building.

One of the aims of the redevelopment was to install a renewable energy technology within the centre. This aim was boosted when it emerged that other local neighbouring organisations shared the desire to increase both their environmental status and their financial viability. It was agreed that a wood-fuelled district heating scheme would be the way forward.

With financial support from the Scottish Government's CARES fund and from South Lanarkshire Rural Capital Fund, and with help and advice from Community Energy Scotland, the scheme was developed. The system is quite large so that it can heat 3 buildings. There are

**St Bride's Centre, Douglas, South Lanarkshire.**
(photograph: Sarah Peters)

Part 2

two fuel stores at either end that can hold 31 tonnes of wood pellets or 12 tonnes of chips in each. The pellets are delivered to the boilers by an auger system.

There are two Guntamatic boilers, one 75kW and the other 100kW, the wood pellets are augered into a chamber where an electric probe ignites the material. Hot flue gases are then passed through a heat exchanger, so that heat is transferred from the combustion chamber. This heats a 5000 litre buffer tank that supplies the hot water to the heating systems ensuring that there is always a supply of heat. The system uses two boilers so that peak demands can be met. The lead boiler changes over periodically to prolong the life of the system.

A separate trading company, Douglas Community Ecoheat was set up to manage the system and supply heat on a not-for-profit basis to the three distinct community buildings – the St Bride's Centre, the Douglas Victoria Bowling Club and the Douglas Valley Church. The heat supplied to each of these buildings is measured separately and each one is billed for the heat it uses.

It's estimated that the total annual financial saving is around £5,000 and the total annual $CO_2$ savings are around 50 tonnes meaning that the total lifetime $CO_2$ savings are expected to be around 1200 tonnes. The system is also used as an educational facility with the local primary school using it to increase their pupils' interest and knowledge of renewable technologies.

Chapter 10

# ANAEROBIC DIGESTION

There is something appealing about energy generated from the power of the wind, the heat and light of the sun, the flow of a river, the movement of the waves or timber from a forest.

By contrast, energy generated from the contents of our wheelie bins, farmyard slurry (liquid manure) or sewage sludge is, well, not so attractive. Renewable energy is supposed to be clean, after all.

However, just because energy from waste is the unglamorous member of the renewable-energy family, that's no reason to ignore its potential. An "energy from waste" project may be ideal for certain communities.

Energy can be generated from different types of waste in a number of ways, but not all of these can really be considered "renewable" forms of energy, and few will be suitable for a community project.

One way is the large-scale incineration of municipal rubbish – the stuff collected by, or on behalf of, a local authority. This is what most people think of when they hear the term "energy from waste". Incineration can be a process that brings environmental benefits, but only if two conditions are met.

The first condition is that as much recyclable material as possible has been removed from the waste before it is incinerated. This includes metals, glass, organic material and some of the plastic. The second is that the heat produced by the incineration process is either used to generate electricity or otherwise usefully exploited – in a district heating system, for example. To fail to make use of the heat is simply a waste.

Is it "renewable" energy, though? It's difficult to think of the burning of our discarded plastic bags, chip-wrappers, polystyrene coffee cups and nappies as "renewable" in the same way as the wind or sunlight, so the answer is probably not. It's also not "low-carbon", as even the most efficient incinerators produce $CO_2$.

Perhaps the only forms of energy from waste that are renewable are those based on biodegradable material – plant or animal material of various types – such as anaerobic digestion (AD) or biogas. The government certainly agrees, as energy produced from organic waste qualifies for payments under the same support schemes (the Feed-in Tariff and Contracts for Difference) as wind and solar, for example.

# Anaerobic digestion

Anaerobic digestion (AD) is one of the processes by which plant or animal matter breaks down – or rots (the other is aerobic digestion, which is what happens with most garden compost heaps). Unlike aerobic digestion, anaerobic digestion produces a gas with high methane content. The rotting takes place in a closed vessel and in a controlled atmosphere. The methane given off is captured and burned to produce heat, electricity or a combination of the two. After a while, the rotting process is more or less complete and you are left with a safe, "clean" organic material that can be used as fertiliser or soil conditioner. AD technology is well-established and is widely used by sewage treatment works to generate electricity. One such plant in Staffordshire, for example, produces enough electricity for themselves with plenty left over to sell to the national grid.

Another example of a working AD plant is at Holsworthy in Devon. This takes hundreds of tonnes of biodegradable waste a week from food processors, abattoirs, supermarkets and biodiesel manufacturers – as well as food waste collected by local authorities. The gas captured during the digestion is used to generate electricity. As long as the supply of waste is relatively stable, the Holsworthy biogas plant can produce enough power for around 8,000 homes.

Is this a model for a community project, though? Probably not, given the logistical, legal and financial complexity of the operation.

There are, however, some communities in rural areas for whom anaerobic digestion – albeit on a smaller scale

– might be ideal. This is particularly the case for groups of dairy or pig farmers who have slurry to dispose of. They could club together – joined perhaps by other local businesses with organic waste to dispose of, like bakeries or cheese processors – and collectively build an anaerobic digestion plant at a suitable site.

All members of the group could deliver their slurry to the site, thereby reducing their mounting costs of storage and disposal. At the same time the plant would generate electricity (and possibly heat), providing income for the group. And, as an additional bonus, the material left over after the anaerobic digestion process could be spread on the land as a soil improver, allowing the farmers to cut down on the purchase of expensive and carbon-intensive artificial fertilisers.

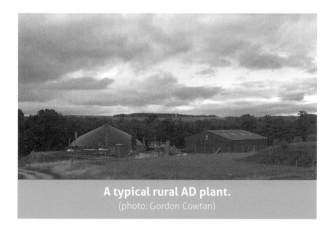

**A typical rural AD plant.**
(photo: Gordon Cowtan)

Although not as controversial as wind, AD plants can cause controversy. For many plants, at least some of the waste will be being brought in using large trucks, which might not be popular with local residents if the plant is in a rural area. There is also some concern about the source of the material being used, with some crops being

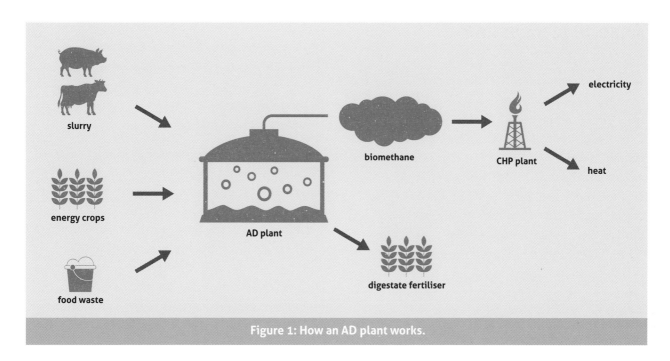

**Figure 1: How an AD plant works.**

grown specifically for use in AD plants, displacing food crops that could be grown in the same ground.

## What about cost?

AD systems don't come cheap – typically a million pounds and over. However, they do have the potential to generate income and make cost savings. Although recent reductions in government funding for renewable electricity have affected anaerobic digesters it's still a prospect rural communities shouldn't shy away from.

Anaerobic digestion is probably the only waste-to-energy technology that a community could look at getting involved in. And if you're interested in this, it's probably

because you're an agricultural community with a large number of dairy farms or pig farms in the area.

## An anaerobic digestion project

Anaerobic digestion, as a form of waste disposal only, is not new and can work at a very small scale. The digester is basically a large tank that breaks down organic materials in an oxygen-free environment, producing methane gas and a liquid, called the digestate, that can be used as a fertiliser. If you want to generate significant amounts of energy as well, then you'll need to add organic materials that have not already been digested, such as grass silage or cereal-crop by-products like straw to the mix.

For many years, the economics of setting up anaerobic digesters for energy production was not very favourable in the UK, and as a result there were very few anaerobic digesters operating at scale.

The government's introduction of the Feed-in Tariff made a significant difference to this, so larger dairy farms find this a somewhat attractive investment, down as low as 350kW-capacity systems, but the best economics are still to be found at a scale that requires more slurry and silage than even the largest farms can produce. As a result, it's probably better to consider anaerobic digestion as a co-operative endeavour to get the best returns.

Really attractive economics can be achieved with plants of about 1MW capacity and above.

## How much slurry and other organic material do we need to produce?

To keep a 1MW anaerobic digester plant going, you'd need about 30,000 tonnes of organic material each year. Some of this has got to be slurry, because it's the bacteria in the slurry that makes the process work, but much of the energy has already been extracted from slurry by the cows that ate it, so you'll need to add other materials too to get the best energy outputs. Grass silage and cereal wastes are great for this, as are fats and oils, but, as in our own diets, not too much fat, please!

You could work on about a third each of slurry; grass and cereal silage; and other material such as waste from chicken farms or cheese processors. While you could actually get away with less slurry, these sorts of quantities would optimise the environmental and economic benefits to your farming community. You would be helping to reduce the amount of methane, a powerful greenhouse gas, emitted from slurry. At the same time you will be creating a valuable soil improver, and helping to deal with the excessive costs of slurry storage.

Given that the average cow produces just over 17 tonnes of slurry per year, you need a group of farmers which have between them around 530 head of cattle to run a 1MW plant, as well as enough land to produce the silage.

## How much energy would we produce?

Such a plant would operate almost all the time, providing a regular supply of enough electricity for about 1,500 homes, which you would feed into the grid. Importantly, it would also generate heat, which could be used on site or distributed via a district heating network. Thinking through how to use this waste heat will be an important part of your plan, and could help you secure finance, as there may be businesses with large heat demands that would be willing to invest in your plant in return for significantly reduced heat costs – horticultural nurseries or bakeries for example.

## Planning permissions and consents

You obviously need planning permission for the plant itself. For this, you will need to provide the local authority with details not only of the plant's structure and design, but also on how you intend to mitigate potential "nuisance issues" such as smell and increased transport.

The planning department will probably require an environmental statement to assess impacts on local eco-systems, and it's very likely that you'll also need to discuss your plans in some detail with the highways department. Anaerobic digesters are increasingly common in the UK and so many planning authorities now have experience of dealing with them and should be able to provide guidance on the information they'll need. However, be prepared as well for some misunderstanding among the local community, particularly confusion between digestion and incineration.

You'll also need various permits from the Environment Agency (or SEPA in Scotland) to operate the plant itself, and to cover the use of the biogas and the digestate. Environment Agency or SEPA permitting can seem completely impenetrable, but it's really a question of ensuring that they have properly logged all waste streams and that watercourses and sensitive environments are not being put at risk.

Don't be put off by the apparent complexity - it's actually quite a standardised procedure - but do make sure you enter into discussions with these organisations early on, as acquiring these permits can be quite time-consuming.

## Costs and income generation

The capital costs of an AD development will vary. One factor in cost is ground type and ease of excavation. This is because, where possible, the tanks are built partially or completely below ground in order to minimise visual impact and stabilise temperatures. Where the water table is very high, the tanks may have to be above ground and well insulated, which can be more costly. A 1MW plant

would probably cost in the region of £3–4m to develop and install.

While you need to take into account the capital costs of the equipment and costs of getting planning permission, you need to offset against that the various other expenses that you will no longer have to pay as a result of using AD for energy production. You should calculate such costs as accurately as possible because these "avoided costs" form a significant part of the overall economics of the system.

Firstly the digestate can be used as a fertiliser, which reduces the cost of buying synthetic fertiliser.

Secondly producing electricity through anaerobic digestion will create waste heat – you can utilise this to heat barns and other farm buildings, or even supply heat to neighbouring properties.

And finally, the costs of disposing of or storing slurry can be very high. If your farm is in a Nitrate Vulnerable Zone (NVZ) then you are not allowed to spread slurry on the land all winter. Providing 26 weeks of slurry storage on such farms can be very expensive.

All of these costs could be significantly reduced through the use of anaerobic digestion for energy production. In addition to these avoided costs, your organisation will receive income – payments from selling electricity to the grid, additional payments from the FIT, and potentially some income from selling metered heat to neighbouring properties.

After the capital installation costs, the main cost to the farmers involved will be the transport of slurry to the digester, and the collection of their proportion of the digestate. For this reason, transport distances should be minimised – five miles or less if possible.

The annual operating income for a 1MW plant, assuming that the majority of the heat is usefully employed, could be between £800,000 and £1m.

## How long will this take?

It is difficult to generalise, but the whole process – from exploratory discussions and meetings through to gaining planning permission, and on to completing the installation – is likely to take in excess of two years.

## What can we do with the digestate?

The process produces about 10% biogas and about 90% digestate – so if you put 10,000 tonnes of slurry in, you will get 9,000 tonnes of digestate back. What you have though, is a much more useful product. Digestate is much less smelly than slurry and is less solid, meaning the land absorbs it more quickly, reducing the time that you can't graze cattle on those fields. In addition, because of the temperatures reached in the tank, weeds and diseases such as bovine TB are killed off so it can considerably improve farm hygiene.

There are still regulations to be applied though. The Environment Agency and SEPA issue permits for the use of digestate as a fertiliser. Speak to them early, and make sure you differentiate between digestate to be spread on your own land and that which you might sell to others; the permit regime is not necessarily the same.

## AD, FITs, CFD and ROCs

Like wind and hydro, the position with regard to financial support for AD projects can be a bit confusing, with FIT, CFD and ROC all being applicable, depending on the size of the project and the date it will be commissioned. However, in reality it's relatively straightforward, with FITs being the best mechanism for any project below 5MW in installed capacity, which is going to account for the vast majority, if not all, AD projects that communities might consider.

Projects above 5MW are eligible for a CFD unless they're commissioned by March 2017, which is unlikely for any project that's not already well under way.

More information about how FITs work in practice together with more information about the CFD and ROC is in Chapter 12.

# Case study of a community owned AD

As I've mentioned a few times in this chapter, for a community group to build and run an AD plant of a sufficient scale to be profitable is difficult and, as if to prove this, there are currently none in the UK that I'm aware of, although there have been a number of attempts over the years and there are still a number of AD projects that communities around the country are working on.

The difficulties that proposed AD projects encounter include:

- Small or micro AD doesn't really exist yet as a technology. Therefore the amounts of money are high: typically a few million pounds is required to build the plant, which is similar in cost to the larger community hydro and wind projects. Any project of this scale involving that much money will be inherently difficult for a volunteer-based organisation to deal with.

- The logistics are difficult. As well as having to deal with planning consents, landowners and grid connections, like many other projects, for an AD plant a community group also has to establish long-term sources of waste for the raw material and long-term agreements for the digestate. And as well as having complex logistics in setting up, AD plants require continual management to run successfully –

much more than similar-sized wind or hydro projects do.

- Few community groups start off with any of the key assets required to make an AD project work. Key assets for an AD plant would typically be suitable land to build it on and owning a reliable source of waste.

- Once introduced to the idea by a community group, landowners can decide to run the project for themselves – after all, they'll often be providing the land and the raw material so the role of the community group can be minimal, and from a commercial point of view, the necessity for their involvement often doesn't exist.

Chapter 11

# COMMUNITY ENERGY REDUCTION

Reducing energy used in a community is of course just as important as establishing new ways of generating it. Energy reduction projects are a good thing to do, not just because they reduce the amount of energy being used, but because they can also bring significant financial benefits to a community – particularly to people who are in fuel poverty and have problems heating their homes – and provide a focus for community activities.

Community energy reduction projects typically involve at least one of three activities:

- Energy advice and home energy audits – performing a study of the amount of energy people in your community are using and offering free advice on measures householders can take.

- Insulation and retrofit programmes – ranging from improving the insulation of a single community building to carrying out a street-by-street home insulation project.

- Behaviour change – changing the way people in your community think about how they use energy (and hopefully, start to use less).

Very often all of these are wrapped up in a single project, which has the aim of creating a shift in the values, beliefs, attitudes and behaviours of people in your community.

## Energy advice and home energy audits

Energy advice and home energy audits often form a part of bigger community energy projects. Run by an organisation that's a trusted source of information in a local community – that understands local issues and who to target – these activities can lead to reductions in energy use and financial savings for local people, and can also help to generate demand for local services like draughtproofing, insulation installation and retrofitting of more carbon-efficient forms of heating like heat pumps.

Energy advice can be a key contribution to a behaviour-change strategy because it alerts and informs householders about their energy use. It might be expected that householders will already know what their energy use is but for many people, thinking about how they can reduce their energy use isn't something they often consider.

There are a wide range of different approaches that can be taken for providing energy advice: for example, running drop-in café sessions on a Saturday morning or one-off sessions in a village hall focusing on specific themes, or setting up a website and producing leaflets providing specific information.

## Home energy audits

The approach to energy advice that can have the biggest impact is to have "energy doctors" or "energy advisors" going door-to-door in a community, compiling information about the use of energy in individual households while at the same time providing free advice on the steps the householder can take to reduce their energy use.

This works even better if you're able to undertake some of the practical steps as part of the same process. For example, arranging for someone to come round and improve the draughtproofing of the house or install insulation in the attic. This way, householders will see the activity as being of direct benefit to them, not just a "talking shop" and will attach more credence to other advice and suggestions you give them.

The information gathered as part of this type of activity is enormously valuable as well. It allows accurate calculations to be made of the different types of energy use within a community and what easy steps might result in reductions being made. The calculations made based on an energy audit can also form the baseline from which other activities can be measured in the future.

Exactly what information is gathered in an audit depends on what you want to do with it, but it would typically include:

- The amount and the types of fuel used over a year – electricity, gas, oil, coal, wood.

- The level and type of transport used by the householders.

- Information about the householders – number and age ranges.

- Information about the house itself – number of rooms, type of heating, levels of insulation.

The information gathered can be held centrally, in a database, as a source of information for future projects.

The success of home energy audits obviously depends on the willingness of householders to provide information. Typically, householders might be reluctant because of the time it will take or because they're being asked to provide information they consider private. There are various ways of overcoming this reluctance. Some projects provide incentives for householders, like free insulation or free energy meters (both these techniques clearly have other benefits as well). It's obviously also important that projects are clear in what they want to do with the information and are able to answer questions about how the data will be stored and who will have access to it (and comply with all regulations about the storage of personal data, including the Data Protection Act).

There may well be a trade-off to be made between getting in-depth information and the number of households taking part in an audit so the more superficial the audit, the more households will take part.

Ideally, home energy audits are not one-off events but are continually monitored, so that changes in energy use are known. How often you want to perform the audit will depend on a wide range of factors, but will rarely be more than annual and often won't be that frequently. Performing audits regularly requires tact on the part of the project so that cooperation from householders can be retained and audit-exhaustion (or perhaps audit-exasperation) doesn't set-in.

## Practical steps

Regardless what approach you want to take with regards to energy advice, it's important to be practical about what you can achieve given the resources that you have. For example, running something like a home energy audit programme on a purely voluntary basis is going to exhaust most community groups very quickly, so be clear about:

- Who you're going to target and how you're going to do this.

- What the realistic outcomes are and what resources you have to achieve this.

It can be a good idea to work with other community groups whose members might have an interest in what you're aiming to achieve. It can also be worth discussing your plans with other organisations in your area, such as the local council. They might want to get involved or they might be able to provide information and resources that are useful.

It's also worth checking that there isn't anyone already providing what you're doing in the local area or if someone has done something very similar nearby. If there is, it might be that you can join forces or learn from their experiences (there are few things quite as good as learning from others' mistakes).

Finally, allow space in your plans for changes. It may be that you find running a drop-in café on a Saturday morning isn't the best approach but that Saturday afternoon works better.

## Case study: Climate Friendly Bradford-on-Avon[1]

Climate Friendly Bradford-on-Avon is a highly active group of volunteers that was founded in May 2006. Within the local community, its aims are:

- Reducing carbon emissions.

- Promoting a greener way of life.

- Raising awareness of climate change.

Over the years, its members have undertaken a wide range of activities, both fun and serious, and have action groups covering energy, waste, travel, biodiversity, food and community engagement. They've defined their own 2050 Carbon Neutral Declaration, which commits signatories to make the local area carbon-neutral by 2050 and to have achieved a target of 40% reduction by 2020. More than 700 people in the local area have now signed this declaration.

One of the activities they ran was a Green Streets[2] project in 2010/11, which incorporated a significant energy audit component. The Green Streets programme provided £140,000 for the project, £100,000 of which was used for household energy audits and advice.

When the project was launched, 135 households initially expressed an interest in taking part. Energy assessments were then carried out in 105 homes. The assessments looked at the householders' lifestyle and energy behaviour, the type and build of the house itself, and did some initial analysis to identify possible renewable energy options.

**Members of Climate Friendly Bradford-on-Avon visiting a biogas plant** (photo: Rowena Quantrill)

The project used an "energy-saving hierarchy" in the recommendations made.

- Starting with Be Lean – cutting energy waste from the lifestyle; then

- Be Mean – making investments in the house that will reduce demand; and finally

- Be Clean – offseting remaining consumption with renewable energy systems.

Typical recommendations from the process therefore ranged from low cost (e.g. energy saving light bulbs and loft insulation) to high cost (e.g. improved boilers and controls, and solid wall insulation) and finally renewable installations like solar PV panels.

Ultimately 98 households undertook some sort of measures, mainly relatively low cost but including 13 boilers, 2 solar PV and 3 solar thermal installations, and also 10 secondary glazing and 5 draughtproofing installations.

As well as this direct impact, there were a number of other benefits, including energy efficiency measures being undertaken by more households than those on the programme and a local carpenter and electrician being employed.

## Insulation and retrofit programmes

Although there's much to criticise in the quality of housing stock currently being built in the UK, particularly with regard to overall sustainability, the levels of insulation and energy efficiency required are now acceptable. Historically, this wasn't the case and in the past 100 years, most homes were built with little or no insulation and with just a nod towards energy efficiency.

Retrofit programmes look to improve the energy efficiency of existing housing. Given that the vast majority of housing stock was built before modern building requirements were in place – the 2010–11 survey of housing in England found that around 87% of homes were built before 1990 – retrofitting energy improvements is clearly of huge importance.

Retrofitting programmes can cover anything from installing draughtproofing to installing a new energy-efficient heating system like a heat pump. They can be very popular in local communities because people see clear benefits from the steps taken – improving the energy efficiency of a home will have financial benefits for the householder and, particularly for people who are in fuel poverty or struggle to heat their homes, can also have significant health benefits as well.

Like energy advice, community groups can run very successful retrofit programmes because they are trusted in the local community and have a sensitivity to what's needed and who to target.

## Insulation

Despite all the reasons given above, in the world of carbon savings – retrofitting and insulation in particular is often seen as the poor cousin. It's not nearly as totemic or exciting to be putting loft insulation into 100 homes or insulating a village hall as it is to be building and operating a wind turbine. The reality, though, is that insulation projects are often easier to get off the ground and will often have a greater, more direct effect on people's lives than big, more dramatic "statement" projects.

There have been a significant number of very successful community-led insulation projects over the years, with some of the most interesting being the door-to-door projects. These projects often work very well at a community level because of the level of understanding and trust that can exist within local communities.

**Installing roof insulation.**
(photo: iStock.com/isturti)

## Practical steps

Because retrofit and insulation projects come in such a wide variety of shapes and sizes, it's very difficult to provide specific advice on the steps to take to run one successfully. Many of the suggestions that apply to energy advice projects also apply to retrofit programmes. In particular, it's important to be practical about what you can achieve, given the resources that you have, so be clear about:

- Who you're going to target, and how you're going to do this.

- What the realistic outcomes are and what resources you have to achieve this.

As a community group you should already be aware of the different types of housing in your local area and, just as importantly, the people who live there and the types of lives they lead – both of these can have a significant effect on what practical retrofit steps can be taken – but don't miss out the obvious step of asking people what they want. Also, make it clear (preferably without exaggerating) what the benefits will be for the house-holders who get involved.

You should find it's possible to negotiate bulk-buy discounts for the measures you are planning – because you'll be buying in bulk, you should have some purchasing power. This applies to almost anything, from insulation through to a reduced cost for drilling bore-holes for ground-source heat pumps.

Over the years there have been a significant number of different government schemes providing financial support for retrofit activities, some better than others. There may well be money you can access to support both your activities as a group and the measures individual householders will be installing.

### Case study: Reepham Insulation Project

Reepham has already been profiled briefly in Chapter 1. It's a market town in Norfolk, with a population of approximately 2,600 living in around 1,200 homes. In 2004 the Reepham Green Team[4] was established which has the aim of making the town a low carbon community.

One of the Green Team's first actions was to undertake a comprehensive community carbon audit. A significant conclusion of this audit was that only 8% of the houses in the town were properly insulated. To tackle this, the Reepham Insulation Project, known as RIP CO2, was set up.

The Green Team decided to deliver this project by working with the local Rotary Club, many of whose members were well-known in the community and were respected and trusted by local residents. It also helped that many of the Rotary Club members were retired so were in a position to be able to help.

The team identified a local company, Anglian Insulations, as the preferred installer and negotiated a discount on their services for local residents.

Two RIP CO2 days were held in April 2009. On each of these days a market stall was staffed by Rotary Club members who had information about the insulation, the benefits and the grant support that was available. Householders who were interested left their contact details and this information was then passed on to Anglian Insulations, which subsequently contacted the householders individually.

This process resulted in 20% of the homes in the town signing up and leaving their contact details, and around half of these homes actually being insulated.

## Behaviour change

Finally in this chapter, there's behaviour change. While insulation is often the Cinderella of energy reduction, behaviour change is the Holy Grail. Behaviour change means persuading people by whatever means possible to alter the way they *think* about energy and therefore modify the way they *use* energy. Successful behaviour change projects should have a huge effect on how much energy people use. This effect can be long-standing and ripple out to people not directly targeted by the project.

If energy generation and energy reduction projects come in a wide variety of shapes and sizes, there's almost no limit to the different forms that behaviour-change projects can take. Because the use of energy is such a significant component of so many aspects of our lives, behaviour change can involve anything from growing food, encouraging bike use, community transport, community-energy workshops and drop-in centres, bulk-buy schemes, energy descent plans, and so on.

This scope is exemplified by the range of projects supported by the Scottish government's Climate Challenge Fund (see page 41). This scheme has been running since 2008 and at the time of writing has supported over 500 projects, the vast majority of which include behaviour change as a key element. These projects include:

- The installation of energy-efficiency measures in community-owned buildings.

- Advice and support on energy efficiency.

- Initiatives to provide lower carbon travel alternatives.

- The provision of community growing space and eating more sustainably.

- Schemes to reduce, reuse and recycle waste.

However, it's fair to say that behaviour change is a very difficult thing to achieve. We live in a world where consumption and even over-consumption is not just acceptable but also aspirational. Behaviour-change projects can find themselves as small voices fighting against this message and small gains can quickly be lost in the maelstrom of modern life. It's therefore important to be realistic about what you can achieve with a behaviour-change project and how you're going to do it.

Because you're often having to battle against the prevailing wind, it's a good idea to keep a behaviour-change project focused on a particular activity and try to tap into some other un-met need within a community, like providing a space for people to grow their own fruit and vegetables in a community where growing spaces aren't already provided.

## Case study: Velocity Café and Bicycle Workshop

As it says on their own website,[5] the Velocity Café and Bicycle Workshop in Inverness is "a social enterprise working to promote healthy, happy lifestyles through cycling". The enterprise started in 2012, providing soup and running bike maintenance courses in a park in Inverness in 2012. They opened their city centre premises in October of that year and now have three strands to what they do:

- A café where you can enjoy local food, hand-baked cakes and soup.

- A fully equipped bike workshop where you can book a workstand to fix your own bike or work with one of the workshops mechanics to learn new skills and leave with a repaired bike.

- The Go ByCycle outreach team who work with other organisations in the local area to provide maintenance classes, safer routes workshops and confidence-building rides.

**Velocity Cafe, Inverness.**
(photo: Laura Nicholson)

Ultimately, the aim of all of these activities is to get more people riding bikes, improving their health and reducing the amount of car use.

The Go ByCycle outreach project was funded for two years by the Scottish Government's Climate Challenge Fund. This funding enabled the organisation to:

- Rent premises.

- Produce promotional material.

- Buy the tools for the workshop which are used to teach bike maintenance.

- Run cycle courses for workplaces and families and run bikeability courses.

The project's long-term aims were to create a modal shift in transport use from car to bike, increase the comm–unity's confidence and knowledge of both cycling and bike maintenance skills, and create a self-sustaining enterprise that can carry on after the funding has finished; plus the project was looking to save 235 tonnes of $CO_2$e over its lifetime.

## Case study: Carbon Conversations

Carbon Conversations[6] is one of the best-known behaviour-change initiatives in the UK. It tackles energy use and carbon emissions head-on and emerged out of an understanding that the factors that stop people engaging with difficult subjects like climate change tend to be psychological and social rather than practical.

Carbon Conversations works by providing a supportive group experience to help people reduce their personal carbon footprint. These groups of 6–8 people meet with trained facilitators in homes, community centres, workplaces or other venues. The meetings are intended to create a non-judgemental atmosphere where people are encouraged to make serious lifestyle changes.

The aim of these meetings is to tackle the difficulties of change by connecting to values, emotions and identity. Materials are provided to cover climate change basics, ideas for a low-carbon future and the four key areas of a carbon footprint – home energy, travel, food and other consumption. Discussions of practicalities are woven together with discussions of how people feel and what these changes mean personally.

It's claimed that carbon reductions of 1 tonne of $CO_2$e are typically made by each member during the course, with plans developed to halve individual carbon foot-prints over a period of 4-5 years. Carbon Conversations is often used to complement other carbon-reduction initiatives and community projects.

Chapter 12

# HOW THE FEED-IN TARIFF AND RENEWABLE HEAT INCENTIVE WORK

This chapter describes how the Feed-in Tariff (FIT) and Renewable Heat Incentive (RHI) work in practice, with information about how to make sure your project complies with the requirements. However, details of how they work are changing all the time, so you should definitely also look at the Ofgem website to get the latest information.

Over the years there have been a number of different mechanisms that the UK Government has put in place to support the renewable energy installations. The history of them is described in more detail in Chapters 3 and 4. All these mechanisms have worked on the same principle of paying the owners of installations an amount for every unit of renewable electricity or heat energy generated.

The most relevant of these mechanisms for community energy projects are the FIT and the RHI. The FIT applies to technologies that generate electricity, like solar PV and wind turbines, while the RHI applies to technologies that directly generate heat, like heat pumps and biomass boilers.

## How do FITs work?

As stated above, FITs apply to:

- Solar PV.

- Wind turbines.

- Hydroelectric schemes.

- Anaerobic digestion plants.

FITs work in largely the same way regardless of the technology although there are significant differences in the amount of money you'll receive.

FIT payments are made by the electricity supply companies and they're also responsible for the application process. Note that not all electricity companies make FIT payments and some are not obliged to accept applications from certain people, so before applying check that your local electricity company will accept your application. Electricity companies that take part in the scheme are known as FIT licensees.

In order to be eligible for FIT payments, there are a considerable number of conditions that your installation must meet. Exactly what conditions apply will vary depending on the technology you're installing and on the installed capacity:

- The total installed capacity of the installation must by 5MW or less.

- If the installed capacity is 50kW or less, then the scheme will go through the MCS route of accreditation. If it's greater than that, then it will go through what's called the ROO-FIT process (see below).

## MCS accreditation requirements

- The equipment must be approved by MCS and the installer must be MCS-certified.

- The installation must be registered on the MCS database within 10 working days of the date it was commissioned.

- Apply for the FIT using the process your FIT licensee provides.

- Once your FIT licensee has checked your application it will provide you with a Statement of FIT Terms, which you must sign and return.

## The ROO-FIT process

For larger installations that need to go through the ROO-FIT process, things are a bit more complicated but essentially there is a two-stage process.

First you need to get the installation accredited by setting up an account on Ofgem's Renewables and CHP

Register and then applying for a new accreditation, which itself can take two different routes: "Full ROO-FIT accreditation" and "FIT preliminary accreditation". You'll need to take the first of these routes if the installation has already been commissioned. The second one is available if the installation has still to be commissioned but allows you to guarantee the FIT rate for the system.

When applying for accreditation for the system, you'll need to provide all the details of the installation, including its location, the date it was or will be commissioned, the total installed capacity, any public grants that were used to help pay for it, information about the metering, and a schematic drawing of the electrical layout.

Once the application has been submitted it will go through two or three stages of review, depending on how complex it is. This can take up to 12 weeks if everything goes smoothly.

The second part of the process happens after your application to have the installation accredited has been approved. You'll receive confirmation by email and you can then approach your chosen FIT licensee that will make the FIT payments.

## After acceptance

Once your installation has been accepted onto the scheme, this is how it works:

- Every quarter (or more frequently for larger installations) you'll send your FIT licensee meter readings for your installation (the need to do this will increasingly be met by the use of smart meters in the future).

- You'll receive a payment based on the amount of electricity produced by your installation plus an additional amount based on the amount that's been exported from your site to the grid (called the export tariff).

- The amount you receive for each kilowatt hour of electricity produced depends on the size of the installation (and also how well insulated your building is). See table 1.

- As described in Chapter 4, the FIT rates are reviewed and adjusted by the government regularly under a scheme called "degression". However, once you're accepted into the scheme the amount you receive is fixed (though adjusted for inflation every April).

- You'll receive the payments every quarter for 20 years after you've been accepted to the scheme.

- If you received a "public grant" towards the cost of the installation of the panels, then this will be deducted from the FIT payments.

- Each year of the scheme you will need to sign a declaration stating that your system still complies with the regulations.

## Solar PV installations

There are different FIT rates based on the installed capacity of the panels. As mentioned above, these are reduced regularly so check on the Ofgem website to see what the current figures are:

| Total installed capacity (kW) | Tariff rate | Export tariff |
|---|---|---|
| 10kW or less | 4.39p/kWh | 4.85p/kWh |
| More than 10kW and less than 50kW | 4.59p/kWh | 4.85p/kWh |
| More than 50kW and less than 250kW | 2.70p/kWh | 4.85p/kWh |
| More than 250kW and less than 1MW | 2.27p/kWh | 4.85p/kWh |
| More than 1MW | 0.87p/kWh | 4.85p/kWh |
| Table 1: Solar PV FIT and export rates.[1] | | |

## Wind turbine installations

Like solar PV, the amount received for electricity produced by a wind turbine under the FIT scheme varies, depending on the capacity of the turbine. As mentioned above, these are reduced regularly so check on the Ofgem website to see what the current figures are:

| Total installed capacity (kW) | Tariff rate | Export tariff |
|---|---|---|
| 100kW or less | 8.54p/kWh | 4.85p/kWh |
| More than 100kW and less than 1.5MW | 5.46p/kWh | 4.85p/kWh |
| More than 1.5MW | 0.86p/kWh | 4.85p/kWh |
| Table 2: Wind turbine FIT and export rates.[1] | | |

## Hydroelectric installations

Unlike solar PV and wind turbines, all hydro schemes must go through the ROO-FIT process – there's no "domestic" scheme for hydro.

Like the other technologies, there are different FIT rates based on the installed capacity of your hydro project. As mentioned above, these are reduced regularly so check on the Ofgem website to see what the current figures are:

| Total installed capacity (kW) | Tariff rate | Export tariff |
|---|---|---|
| 100kW or less | 8.54p/kWh | 4.85p/kWh |
| More than 100kW and less than 2MW | 6.14p/kWh | 4.85p/kWh |
| More than 2MW | 4.43p/kWh | 4.85p/kWh |
| Table 3: Hydro FIT and export rates.[1] | | |

## Anaerobic Digestion

Like hydro schemes, because of their nature, all AD projects must go through the ROO-FIT process to sign-up for FiTs.

These FIT rates for AD plants are currently:

| Total installed capacity (kW) | Tariff rate | Export tariff |
|---|---|---|
| 250kW or less | 9.12p/kWh | 4.85p/kWh |
| More than 250kW and less than 500kW | 8.42p/kWh | 4.85p/kWh |
| More than 500kW | 8.68p/kWh | 4.85p/kWh |
| Table 4: AD FIT and export rates.[1] | | |

## Deployment Caps

Following the review of FITs in 2015, as well as there being substantial reductions in the many tariff rates, a new mechanism was introduced called Deployment Caps.[2] A maximum of £100 million per year was set for all new FIT installations. This was divided into quarters between technologies and bands. The result was quarterly maximum installed capacity caps as defined in table 5 on the next page.

If a cap is reached before the end of the quarter, then all subsequent FIT applications are "frozen" and held over in a queue until the next quarter. Any "underspends" are rolled over to the next quarter, although periodically there will be redistributions of underspends as deployment top-ups. For the latest information on how deployment caps work, see the Ofgem website.

Part 2

101

| Technology bands | | 2016 | | | | 2017 | | | | 2018 | | | | |
|---|---|---|---|---|---|---|---|---|---|---|---|---|---|---|
| | | Q1 | Q2 | Q3 | Q4 | Q1 | Q2 | Q3 | Q4 | Q1 | Q2 | Q3 | Q4 | Q1 |
| PV | less than 10kW | 48.4 | 49.6 | 50.6 | 51.7 | 52.8 | 53.8 | 54.2 | 55.9 | 57.0 | 58.0 | 59.1 | 60.1 | 61.1 |
| | 10-50kW | 16.5 | 17.0 | 17.4 | 17.8 | 18.2 | 18.6 | 18.7 | 19.4 | 19.8 | 20.3 | 20.7 | 21.1 | 21.5 |
| | more than 50kW | 14.1 | 14.5 | 14.9 | 15.4 | 15.8 | 16.2 | 16.4 | 17.1 | 17.6 | 18.0 | 18.5 | 19.0 | 19.4 |
| | Standalone | 5.0 | 5.0 | 5.0 | 5.0 | 5.0 | 5.0 | 5.0 | 5.0 | 5.0 | 5.0 | 5.0 | 5.0 | 5.0 |
| Wind | less than 50kW | 5.6 | 5.6 | 5.5 | 5.5 | 5.6 | 5.5 | 5.5 | 5.4 | 5.5 | 5.4 | 5.4 | 5.3 | 5.4 |
| | 50-100kW | 0.3 | 0.3 | 0.3 | 0.3 | 0.3 | 0.3 | 0.3 | 0.3 | 0.3 | 0.3 | 0.3 | 0.3 | 0.3 |
| | 100kW-1.5MW | 6.8 | 6.7 | 6.6 | 6.5 | 6.4 | 6.3 | 6.2 | 6.1 | 6.1 | 5.9 | 5.8 | 5.7 | 5.7 |
| | 1.5MW-5MW | 5.0 | 5.0 | 5.0 | 5.0 | 5.0 | 5.0 | 5.0 | 5.0 | 5.0 | 5.0 | 5.0 | 5.0 | 5.0 |
| Hydro | less than 100kW | 1.1 | 1.1 | 1.2 | 1.2 | 1.2 | 1.3 | 1.3 | 1.3 | 1.3 | 1.3 | 1.4 | 1.4 | 1.4 |
| | 100kW-5MW | 6.1 | 6.2 | 6.3 | 6.3 | 6.4 | 6.4 | 6.4 | 6.4 | 6.4 | 6.4 | 6.4 | 6.3 | 6.3 |
| AD | All | 5.8 | 5.0 | 5.0 | 5.0 | 5.0 | 5.0 | 5.0 | 5.0 | 5.0 | 5.0 | 5.0 | 5.0 | 5.0 |

Table 5: FIT deployment caps.

# How does the RHI work?

As stated earlier, the RHI applies to:

- Solar thermal.

- Ground-source and air-source heat pumps.

- Biomass.

Like FITs, the RHI works in much the same way, regardless of the technology, though the amount of money you receive varies. Also, like FITs, there are a considerable number of conditions that your installation must meet. These conditions depend on the technology to a certain extent, but also on the type of building in which the system is being installed – whether it's an installation in a home or in a community or business building.

## Installations in homes

The conditions that must be met by any installation in a home are:

- Anyone can apply, including homeowners, private landlords and people who've built their own home.

- Your installation must have a Microgeneration Certification Scheme (MCS) certificate. This means the equipment must be MCS-certified and installed by an MCS-certified installer.

- The entire system must be new when commissioned.

- Your application to join the scheme must be within 12 months of the date the system was commissioned.

- You'll need an Energy Performance Certificate (EPC) for the building to ensure it's a "domestic dwelling".

In the past it was also a requirement that you had a Green Deal assessment carried out for your property and any recommended insulation works carried out. Since the closure of the Green Deal, this is no longer the case.

If your installation is eligible, then this is how the scheme works:

- You'll receive a payment for every kilowatt hour of heat energy produced by the system. This is usually based on a "deemed" amount of energy produced*.

- As described in Chapter 4, the amount that will be paid is reviewed and adjusted by the government regularly under a scheme called "degression". However, once your installation is accepted into the scheme the amount you receive is fixed, though adjusted for inflation every April.

- You'll receive the payments every quarter for seven years after your installation has been accepted into the scheme.

---

* For domestic heat installations, metering the amount of heat energy produced would be far too expensive so in calculating the RHI an estimated or "deemed" amount of energy is usually used. For solar thermal the deemed amount is usually calculated by the MCS installer.

- If you received a "public grant" towards the cost of the installation, then this will be deducted from the RHI payments over the seven years.

- Each year of the scheme you will need to sign a declaration stating that your system still complies with the regulations.

There are a number of other rules for the scheme, plus it's being tweaked and updated regularly so you should read the most recent information about the scheme on the Ofgem website.[3]

### Solar hot water

Conditions specific to solar thermal systems include:

- The panels have to be either flat plate or evacuated tube collectors.

- The panels must be being used for heating domestic hot water (so they can't be used for space heating, for example, nor can they be used for swimming pools).

At the moment, you'll receive 19.51p for every kilowatt hour of heat energy produced by the panels.

### Heat pumps

Conditions specific to heat pumps include:

- The system can be either an air-source or ground-source heat pump.

- The system has to be used for space heating (i.e. heating your house) or space heating and water heating (it can't be used for swimming pools).

- It must use a "wet" central heating system like radiators.

- The heat pump must have a minimum seasonal performance factor (SPF) of at least 2.5 (see below for an explanation of what this means).

At the moment, you'll receive 7.42p per kilowatt hour if it's an air-source heat pump or 19.1p per kilowatt hour if it's a ground-source heat pump.

### Calculating RHI payments for heat pumps

Calculating RHI payments for domestic installations without heat meters (which is the vast majority) is not straightforward. There are three key figures – the amount of heat the heat pump has generated, the seasonal performance factor for your heat pump and the tariff rate.

The calculation of how much heat your heat pump is generating is based on the heat load figure for your house in the Energy Performance Certificate (EPC). This is basically an estimate of the amount of heat energy it takes to heat your house.

Heat pumps use electricity to work, so we then need to calculate the renewable portion of the heat pump's output by deducting the amount of electricity used. This is where the Seasonal Performance Factor (SPF) comes in. The SPF is a measure of how efficient your heat pump is over a whole year's use and is defined as the total heat energy output by the heat pump in a year divided by the total amount of electricity used by the heat pump in a year. So, for a heat pump that outputs 2,500kWh of heat in a year and uses 1,000kWh of electricity to do this, the SPF would be 2.5.

Obviously this could be calculated using a meter on the electricity supply to the heat pump, but for the vast majority of domestic heat pump installations it's done using an estimated SPF. All heat pumps installed after 9 April 2014 will have a specific estimated SPF value based on the system's Temperature Star Rating, which is shown on the MCS Compliance Certificate for the system.

For a house with a total heat demand of 10,000kWh and a ground source heat pump with an SPF of 3 the calculation would therefore be as follows (using the current tariff rate of 18.8p):

Total RHI payment $= 10,000 \times (1 - \frac{1}{3}) \times 0.188$

$= 10,000 \times \frac{2}{3} \times 0.188$

$= £1,253.33$

You would then receive a quarter of this every three months.

## Biomass systems

Conditions that apply specifically to biomass systems include:

- The system can be either a biomass boiler that uses solid biomass fuel (e.g. pellets, woodchips or logs) or a biomass stove that uses wood pellets only (e.g. cannot use chips or logs).

- The system must be used for space heating (i.e. heating your house) or space heating and water heating (though it can't be used for heating a swimming pool).

- Some stoves that can also be used for cooking are also eligible.

- It must use a "wet" central heating system like radiators; for a stove, it must use a liquid-filled heat exchanger.

### Calculating RHI payments

At the time of writing, you'll receive 5.14p for every kilowatt hour of heat energy produced by the system.

Calculating RHI payments for domestic installations of biomass systems without heat meters (the vast majority) is fairly straightforward and very similar to the calculation required for heat pumps. There are two key figures – the amount of heat the system has generated and the tariff rate.

Because the system isn't metered, the amount of heat it produces is estimated, based on the heat load figure for your house in the EPC. This is basically an estimate of the amount of heat it takes to heat your house.

The annual heat load is then multiplied by the tariff rate to give your annual RHI payment.

For a house with a total heat demand of 10,000kWh the calculation would therefore be as follows (using the current tariff rate of 5.14p):

Total RHI payment $= 10,000 \times 0.0514$

$= £514$

You would then receive a quarter of this every three months.

## Installations in businesses or community buildings

Renewable heating systems that are not part of homes or houses should be eligible for the Non-Domestic RHI[4]. One of the main differences between the Domestic and Non-Domestic RHI is that the payments under the latter last for 20 years rather than 7.

Like the Domestic RHI, in order to be eligible, there are a lot of conditions that must be met:

- The entire system must be new when commissioned.

- Your application to join the scheme must be within 12 months of the date the system was commissioned.

If your installation is eligible, then this is how the scheme works:

- You'll receive a payment for every kilowatt hour of heat energy produced by the system.

- The amount of heat energy produced must be measured using heat meters and reported every quarter.

- As described in Chapter 4, the RHI rates are reviewed and adjusted by the government regularly under a scheme called "degression". However, once you're accepted into the scheme the amount you receive is fixed, though adjusted for inflation every April.

- You'll receive the payments every quarter for 20 years after you've been accepted to the scheme.

- If you received a "public grant" towards the cost of the installation of the panels, then this will be deducted from the RHI payments.

- Each year of the scheme you will need to sign a declaration stating that your system still complies with the regulations.

### Solar hot water

Conditions specific to solar thermal systems include:

- The panels must be either flat plate or evacuated tube collectors.

- The panels must be used for space, water or process heating.

- The total capacity of the installation must be less than 200kW and if less than 45kW it must have an MCS certificate.

- The panels can't be used on a single home (but could be used as part of a district heating scheme heating water for a number of homes).

At the time of writing, you'll receive 10.16p for every kilowatt hour of heat energy produced by the panels.

### Heat pumps

Conditions specific to heat pump systems include:

- The heat pump can be ground-source, water-source or air-source.

- The heat pump must be used for space, water or process heating within a building or cleaning or drying.

- There's no limit on the total capacity of the installation but if it's less than 45kW it must have an MCS certificate.

- The heat pump can't be used on a single home (but could be used as part of a district heating scheme heating water for a number of homes).

- The installation must include heat metering.

At the time of writing, for ground-source and water-source heat pumps you'll receive 8.84p for every kWh of heat energy produced in tier 1 (see below) and 2.64p for every kWh of heat energy produced in tier 2, while for air-source you'll receive 2.54p for every kWh of heat energy produced.

### Tier 1 and tier 2

Tier 1 and tier 2 are simply references to the total amount of heat produced by your heat pump. Heat energy up to and including 15% of the total capacity over a year is tier 1 and anything else your heat pump produces is tier 2.

The easiest way to work out what this means in practical term is to look at an example. Imagine that your heat pump has a rated output of 50kW. There are 8,760 hours in most years (a bit more in a leap year) so the total your heat pump could theoretically produce in a year is 50 x 8,760, which is equal to 438,000kWh.

So 15% of the total capacity would be 438,000 x 15%, which is 65,700kWh.

This means that if your heat pump produced 100,000kWh of heat during a year, the first 65,700kWh of this would be qualify for the tier 1 RHI rate, while the remaining 34,300kWh of this would qualify for the tier 2 rate.

The reason for separating the heat produced into these two tiers is to reduce the incentive for people to over-generate heat simply to claim additional RHI payments. Currently tier 1 and tier 2 are not applicable to air-source heat pumps.

### Calculating RHI

Because Non-Domestic RHI installations must be metered, calculating the RHI payments is actually much more straightforward than for Domestic installations (despite tier 1 and tier 2).

If we take the same 50kW ground-source heat pump as in the previous section. We know that 65,700kWh of the renewable heat generated is eligible for tier 1 payments and 34,300kWh is eligible for tier 2, while the tier 1 rate is 8.84p and the tier 2 rate is 2.64p.

So, the total annual payment would be:

$$(65,700 \times 0.0884) + (34,300 \times 0.0264)$$

$$= £5,807.88 + £905.52$$

$$= £6,713.40$$

### Biomass systems

Conditions specific to biomass systems include:

- The biomass system can use any type of solid biomass (e.g. logs, chips or pellets) and can also use solid biomass contained in waste.

- The system must be used for space, water, process heating within a building or cleaning or drying.

- There's no limit on the total capacity of the installation but if it's less than 45kW it must have an MCS certificate.

- The system can't be used for a single home (but could be used as part of a district heating scheme heating water for a number of homes).

- The installation must include heat metering.

### Calculating RHI

At the time of writing, small biomass systems with less than 200kW capacity receive 3.76p for every kWh of heat energy produced in tier 1 and 1.0p for every kWh produced in tier 2 (see below for more information on what the different tiers are); medium biomass systems between 200kW and 1MW receive 5.18p for every kWh in tier 1 and 2.24p in tier 2; while large systems of 1MW or above receive 2.03p for every kWh produced.

Tier 1 and Tier 2 for biomass are identical to Tier 1 and Tier 2 for heat pumps i.e. heat energy up to and including 15% of the total capacity over a year is tier 1 and anything else your system produces is tier 2.

Because Non-Domestic RHI installations must be metered, calculating the RHI payments is actually much more straightforward than for Domestic installations (despite tier 1 and tier 2).

If we take the same 100kW biomass system as in the previous section, we know that 131,400kWh of the renewable heat generated is eligible for tier 1 payments and 68,600kWh is eligible for tier 2, while the tier 1 rate is 6.8p and the tier 2 rate is 1.8p.

So the total annual payment would be:

$$(131,400 \times 0.0376) + (68,600 \times 0.010)$$

$$= £4,940.64 + £686.00$$

$$= £5,626.64$$

# Contracts for Difference and Renewable Obligation Certificates

*In 2015, the government announced that on-shore wind energy projects would no longer be included within the Contracts for Difference (CFD) scheme.*

The Contract for Difference (or CFD – though just to confuse things a bit more, it should be called FIT-CFD) is the mechanism for supporting low-carbon electricity generation introduced as part of a package of energy market reforms. The CFD lasts for 15 years and is only applicable to projects of 5MW installed capacity or bigger.

The aim of the CFD is to provide long-term contracts, which will give power companies a guaranteed price for the low-carbon electricity they produce. This is intended to reduce the risk of investment in projects with high upfront capital costs. The CFD applies to other forms of electricity generation as well as renewables, including carbon capture and, controversially, nuclear reactors.

This section is relatively short because CFDs are only relevant for large projects over 5MW, which are rare in the world of community renewables (the obstacles for a community or group of communities wanting to develop a project of this scale are considerable). Added to this, in 2015, the government announced that on-shore wind energy would no longer be included with the CFD scheme and at the moment, it's unclear whether this policy change is permanent or will change again.

What's more likely than a community instigating and owning a development of this is size is that a community might have a share in a large commercial wind farm, in which case it's important that the community group understand how the income from the wind farm is generated.

The CFD works in tandem with the wholesale energy market. The government sets a predetermined price, called the "strike price", for renewable generation – see Table 6 on the next page. Generators (owners of wind farms and other renewable generation) then sell the electricity they produce to suppliers at the market wholesale price. If this market price falls below the strike price, it is topped up by an extra payment and if the market price is above the strike price, generators pay back the surplus.

The CFD is allocated on a regular basis by the government holding auctions with a limited budget to allocate and maximum and minimum strike prices, and developers submitting sealed bids for their developments. The allocation process then works as follows:

- All projects would first be ranked by their bid, representing the strike price they would be willing to accept.

- Those projects that can offer the lowest bids would be allocated a CFD first.

- Allocation would then proceed to the next cheapest projects in the group.

- This would continue until the entire budget allocated to the group had been used up.

At the time of writing, the first of these auctions is taking place with a maximum strike price of £95 per MWh.

It should be obvious that compared with FITs, the CFD is a very complex mechanism to get involved with. There is a clear risk that if the price you bid for your project is too high, then you will fail at the auction and your whole project will fail. The government has published maximum strike prices for the next few years (see Table 6 below). If you're thinking of developing a large project, it's vital that you get professional advice and financial modelling on the income required for it to be viable.

| Year | Strike price £/MWh |
|---|---|
| 2014/15 | 95.00 |
| 2015/16 | 95.00 |
| 2016/17 | 95.00 |
| 2017/18 | 90.00 |
| 2018/19 | 90.00 |

Table 6: CFD strike prices.[5]

## The Renewable Obligation Certificate (ROC)

The CFD is replacing the Renewable Obligation Certificate (ROC), which were the support mechanism introduced in 2002 that had been used for larger projects until 2015 (in fact, before FITs were introduced, ROCs were the only support mechanism for renewable generation). Because ROCs are being phased out so soon, your project would have to be well under way for them to be an option.

# PART 3
# LEGALS, FINANCIALS AND PLANNING

While Part 2 looked at the practicalities of the technologies and other measures that community energy projects often involve, Part 3 looks at some of the other practical aspects that community energy groups have to think about. Firstly there's a chapter on legal structures that describes the various types of legal entity (wake up at the back there!) that are available to community groups, outlines their good and bad points and also looks at situations where they work best.

There's then a chapter on financial models, particularly with regard to energy generation projects. It looks at how finance for these kinds of projects usually works and walks you through three different examples of increasing complexity, looking at how the cash flow for each of them can be estimated over the lifetime of the project.

Finally, there's a chapter covering a number of the other practical issues that community energy groups have to deal with, like how your group is organised, gaining planning permission for your project, gaining support for your project and raising finance.

Chapter 13

# LEGAL STRUCTURES

Getting "legal stuff" right matters. It's not nearly as exciting as thinking about installing a wind turbine or biomass boiler, but getting the legal structure right for your community energy project matters because it can have a strong influence on the success of your project and has a direct effect on, for example, liability, ownership, funding, governance, profit distribution and charitable status. Not all of these will be relevant to your project, but it's worth thinking through each of them to make sure you're not overlooking something that might later come back and bite you or someone else.

## What's meant by "legal structure"?

The "legal structure" is the legal form your organisation takes. We're all familiar with the notion of a limited company, which is the legal form used by most commercial enterprises. However, in the commercial world as well as limited companies there are also sole traders, partnerships, limited liability partnerships, and so on. Because community energy covers a wide range of different sizes and types of group, there are even more options when it comes to which legal structure is the best fit for what you're wanting to do.

To confuse matters even more, many projects end up using a number of different structures for different aspects of what they're doing. For example, you might have a charitable organisation that carries out the main charitable activities of your group but which also owns a trading company that operates a profit-making wind turbine, which in turn donates its profits to the charity.

This probably sounds a bit daunting, but it needn't be and, usually, if you've got a clear idea of what you're planning, the correct legal structure to use will be fairly obvious.

## Why do legal structures matter?

The legal structure you use matters because ultimately it can affect almost every aspect of your project. The following are some of the questions and issues you need to think about when considering legal structures.

### Are you going to be a membership organisation?

By the nature of community energy initiatives, most have members. It's how people living in your community have a direct involvement in your project. Membership organisations can take a number of different legal forms but they all have the same principle of having a membership that elects a board of directors or trustees. This board will take the main decisions but are ultimately answerable to this membership.

## Are you planning to raise money and if so how?

The chances are that unless you're going to be an entirely voluntary organisation, you're probably going to need to raise money from somewhere to do some of the things you want to do. How you plan to raise that money – from grants, from individual investors or from loans – can have a significant bearing on your legal structure.

If you do have individual investors, are you planning to distribute profits to them? Organisations that distribute profits (or their assets on dissolution) will not usually be eligible to receive grants either from government bodies or grant-giving trusts. The opposite is also true though: if you want to attract private investment (other than by way of loans), you'll need to have a structure that enables you to distribute profits to those investors.

## Are you going to sell things in some way?

Any community energy project that is going to generate electricity (e.g. by installing PV panels) or heat (e.g. by installing a biomass boiler) is likely to be selling what's generated to someone else. If you're trading in this way, some legal structures will work better than others.

## Liability

It's important to know what's going to happen if something goes wrong. If your project suffers a financial loss, who ends up being responsible for this loss? Many of the legal structures like limited companies or community benefit societies will insulate the directors and members from personal liability for financial losses (though be aware that this protection can be undermined in the case of board members if they fail to comply with their legal duties).

## Ownership

Different legal structures incorporate different ownership structures. If the organisation is a company limited by shares, it's usually the case that the more shares somebody holds, the more of the company they own and the bigger the say they can have in decisions. A company limited by guarantee is based on the notion of membership, with each member normally having a single vote. A co-operative can be considered somewhere between these two, where each member will have a single vote, but the return they receive on their investment will be based on how much they invest.

Issues of ownership are obviously tied up with whether you're planning to raise investment money, how you're planning to do this and for what purpose.

## Governance

Governance is one of those terms that most of us don't come across in everyday life, but it means the decision-making and accountability structures, procedures and systems of an organisation. Basically how does your organisation run, who makes the decisions, how are decision makers held to account, and how open is the organisation in how it operates?

Governance is important for all organisations, but particularly so for organisations that have memberships, are handling "community money" and are aiming to make positive changes for people. It's important that people can see how your organisation is run and can ask questions about decisions.

### Are you eligible to be a charity and do you want to be one?

Being a registered charity can have significant benefits for your organisation in terms of its credibility and access to tax reliefs; it also tends to bring the widest range of potential funding sources. However, there are significant criteria that you have to meet before you can gain charitable status. These can be too restrictive for some community energy projects; also having charitable status adds a significant burden to the upkeep of your organisation in the form of increased governance. This, of course, is a good thing in many ways but is an additional burden that many voluntary organisations decide isn't worth it.

## No legal structure

It might seem a bit contrary – given how important I've said having the correct legal structure is – but, for some projects, not having any formal legal structure at all can be the right option and shouldn't be ruled out. These would typically be very simple ad hoc projects like a group of neighbours getting together to organise a bulk-buy scheme for insulation or heat pumps. If each householder is entering separately into a contract with the supplier and accepts that they are responsible for their own decision then there's no reason that any type of formal organisation is required at all. If one person is

doing all the initial buying though, you might want to think about where the liability will lie if anything goes wrong.

So even if you think that what you're doing really won't need any kind of legal structure, it's still worth going through each of the aspects above to make sure that there are no repercussions either now or later.

(Note that you might find, even in the simple example above, you are unwittingly an "unincorporated association" – see below – through the simple act of associating with others in pursuit of shared goals.)

## Membership organisations

For many community energy projects, some type of membership organisation is the most appropriate form to take.

### Unincorporated associations

Unincorporated associations are widely used in the voluntary sector for any kind of organisation that has members, for example sports clubs and residents' associations. They are the most straightforward type of membership organisation (the term "unincorporated" simply means that the organisation has no legal existence separate from its members – see Liabilities opposite) and many organisations will form as unincorporated associations in their early days because they are relatively easy to set up.

An unincorporated association will usually have a constitution or a set of rules and a management committee, which will be elected to run the association on behalf of its members. However, while they're relatively simple and straightforward, there are some significant downsides of unincorporated associations.

### Liabilities

One of the significant downsides of unincorporated associations is that they are not treated as separate legal entities as far as liabilities are concerned. This means that each member of the management committee (and potentially each member of the association) could be liable for any type of liability the organisation might incur (if the organisation is unable to meet such liabilities out of its own resources). So if you are likely to be doing anything like employing staff, raising significant amounts of money or signing substantial contracts, you should think very carefully about whether an unincorporated association is the right legal form for what you're doing.

### Funding

In practical terms, unincorporated associations can't raise funds via share issues or any similar approach. However, although they don't have a separate legal identity, they can hold bank accounts and are often eligible for grants or even bank loans in the right circumstances. Having said that, many grant bodies might want to see a more robust structure or require the association to have charitable status.

### Governance

As mentioned above, unincorporated associations aren't required to register with Companies House or any other regulatory body – though they are required to keep financial records, file a tax return and (if they don't have charitable status) pay corporation tax. This means that they're generally more flexible and easier to maintain with regard to governance.

### Profit distribution

If you want to run a community energy project that's going to distribute profits in some way, then an unincorporated association is not the way to do it.

### Charitable status

Unincorporated associations can be charities, provided that you can demonstrate that you have charitable aims and that your work is for public benefit. This is discussed in more detail later in the chapter – but if you want to improve the governance and credibility of your unincorporated association, gaining charitable status is one way to do it. Having said that, if you are facing the prospect of taking forward an application for charitable status (which is not straightforward), it would probably be as well to adopt the Charitable Incorporated Organisation (CIO) or, in Scotland, the Scottish Charitable Incorporated Organisation (SCIO) legal form, as it adds very little to the process and gives the benefit of limited liability and a clear legal identity.

### Costs

Setting up an unincorporated association is usually fairly straightforward and inexpensive. There are many template constitutions available on the internet supplied by umbrella organisations.

Part 3

*Summary*

Unincorporated associations work best for relatively small organisations that don't intend to do anything like lease premises or employ staff.

## Company limited by guarantee

We're all familiar with the term "limited company" and a "company limited by guarantee" is a type of limited company. However, rather than having shareholders, which most limited companies have, a company limited by guarantee has members. People (and sometimes organisations) are able to join as members; and the members elect a board of directors, who run the company on behalf of the members. In this regard, a company limited by guarantee is very similar to an unincorporated association.

From a practical point of view, the big differences between the company limited by guarantee and the unincorporated association are:

1.  A company limited by guarantee must be registered with Companies House and must comply with the rules for operating companies within the UK – including submitting annual accounts and annual returns to Companies House.

2.  A company limited by guarantee is a separate legal entity from the members. This means that it is the company itself that enters into contracts and employs people, and the members have limited liability if the company runs into problems. (Note that although members have this protection, directors can still be held personally liable. Anyone who is or is thinking of becoming a director needs to be aware of their duties and it's certainly worth accessing guidance on this aspect.)

A company limited by guarantee has "articles of association" which set out the aims and objectives of the organisation and procedures for membership application, electing directors, running meetings and similar matters. A copy of the articles is lodged with Companies House when the company is formed; and any alterations (similarly, any new set of articles) need to be filed with Companies House.

Importantly for many community energy groups, companies limited by guarantee can own other companies – so they're often used as the main umbrella membership organisation, which then owns other trading organisations. This is discussed in more detail in the section on "Hybrid structures" – see page 123.

*Liabilities*

As mentioned above, a company limited by guarantee provides limited liability for its members – normally liability is limited to a token £1.

*Funding and profit distribution*

If you're looking to raise funding from grants or via a bank loan, a company limited by guarantee is often an appropriate form. However, if you want to raise money from private investors and distribute profits to them, it's not the best approach to take.

## Charitable status

Depending on its activities, a company limited by guarantee can be eligible for charitable status, but its articles of association will have to be approved and it will have to comply with the other rules for being a charity. If you want to be a charity, you may find that the CIO or SCIO legal form works better since it avoids the need to comply with the requirements of both Companies House and the relevant charity regulator simultaneously (see page 123).

## Costs

Setting up a company limited by guarantee is more expensive than an unincorporated association. You will usually need to employ a lawyer to make sure you have the articles of association correct and you'll need to register with Companies House, and so on. The running costs are greater as well, with the need to produce company accounts every year meaning you'll normally need to engage an accountant.

## Summary

The structure of the company limited by guarantee is often used by voluntary organisations that employ staff, regularly enter into contracts, manage investments, or own property and other assets, because limited liability helps to minimise the threat of personal liability.

Many community energy organisations use the form of the company limited by guarantee because it provides the membership structure, limited liability and a degree of flexibility.

## Registered societies

Registered Societies are what many people think of as co-ops. Co-ops are indeed one form that a registered society can take and many of the principles behind co-ops appear in the other form, the BenCom – see the following page.

Registered societies are organisations that are run and owned by their members. A society can operate just for the benefit of its members or for the benefit of the community. Societies are "incorporated" so – like a company limited by guarantee – have legal identities separate from their members and can own property, enter into contracts, issue shares and take out loans.

Part of the definition of a co-operative is

> An autonomous association of persons united voluntarily to meet their common economic, social and cultural needs and aspirations through a jointly owned and democratically controlled enterprise."[1]

This "autonomous association of persons" is essentially the membership of the co-operative. Registered societies generally operate on the basis of open membership and having "one member, one vote" (although for some types of society this isn't a legal requirement).

Like the other legal structures we've looked at, registered societies tend to have a management committee who manage the organisation on behalf of the members, but (as with a company limited by guarantee) ultimately it's the members who have democratic control.

Part 3

## Liabilities

As mentioned above, societies are "incorporated" so have their own legal identity that protects members. If a registered society becomes insolvent, the members will lose whatever money they spent in purchasing shares. In most cases – other than where there is a community share issue – that will be a token £1 paid for one share on becoming a member of the registered society.

## Funding and profit distribution

Societies can raise funding through loans and share issues to individual investors. They can also provide a return on investment (see "Community benefit societies") provided it's within certain constraints.

Unless they have charitable status or have specifically adopted an "asset lock", societies can be more problematic for grant-funding bodies – since there is a perception that there is a higher risk of the members altering the rules so as to pay themselves surplus profits by way of dividends (or even winding up the registered society and distributing the net assets among themselves).

## Governance

The "one member, one vote" principle (which applies in most registered societies) means that societies tend to be open and democratic, in a similar way to companies limited by guarantee. Societies have to be registered with the Financial Conduct Authority (FCA) rather than Companies House, and must submit annual accounts to the FCA.

## Charitable status

There is a particular type of society called a society for the benefit of the community ("BenCom" – see below) that can have charitable status. In a community energy context, this legal form would be the preferred model if there was an intention to take forward a community share issue.

## Costs

As mentioned above, societies have to be registered and must submit annual accounts to the FCA. Obviously if you have issued shares in your society and you have investors, there is also an overhead in managing the investments and communicating with the investors. One way or another there can be a significant overhead to running a society. In addition, the set-up costs can be much more expensive (and potentially more complex) than for a company limited by guarantee, unless the group closely follows a set of model rules that has already been approved by the FCA.

## Community benefit societies (BenComs)

Following an update to the law in 2014, there are now distinct types of society in the UK. From our viewpoint, the most interesting are societies for the benefit of the community (often referred to as BenComs). One of the key features of a BenCom is that it is directed towards providing benefit to the community, unlike the co-op form of registered society, which is focused on providing benefit to its members.

One of the other features of BenComs is that they have the option of adopting an asset lock, which prevents the distribution of residual assets to members. This means that the community benefit is maintained and cannot end up as private benefit to the members.

BenComs therefore combine a number of characteristics that many community energy projects want:

- Limited liability.

- Democratic control on the "one member, one vote" basis.

- The ability to raise investment from private individuals.

- The ability to provide a return on that investment (up to certain limits).

- The ability to distribute benefit outside the membership to the wider community.

- The ability to create an asset lock.

Over the past few years there has been an increasing use of BenComs (primarily in England and Wales, but now gradually taking hold in Scotland) as the legal vehicle of choice for community energy projects that need significant capital to get started – like community hydro or wind projects. BenComs allow community groups to raise funds from local people, give them a return on their investment (even if that return isn't huge) and also distribute profits from the scheme to other projects benefiting the local community.

This is a good thing, but it does have downsides that are to do with the investment. Firstly it's difficult to restrict the location of the investors to a particular part of the country, so schemes can have investors from hundreds of miles away. In reality, most schemes are only publicised locally so most investors are from nearby but a significant number are not.

Secondly not everyone locally will be able to invest in the scheme; and for many schemes the majority of the income generated – for example from the sale of electricity – goes to the investors rather than to fund other community projects. The result of this is that some people within a community will benefit from it far more than others. There may also be concerns over the risk of investors adjusting the rules in the future, so as to increase their personal financial return – since the power to alter the rules sits with the investors as the members of the BenCom.

Thirdly the model rules for BenComs currently used in the context of community share issues do not tie the BenCom in with existing community structures. Often, for example, there will be an existing community development trust – and it may be unhelpful to have a completely independent BenCom, with its own body of members, which may or may not be aligned with the overall strategies set by the development trust.

Unfortunately, there is no ready-made solution to these problems: the capital for the projects needs to come from somewhere and having most of it from local people is better than many of the other options. However, I remain wary that raising investment through BenComs seems to be the default route and is starting to become increasingly popular with community energy generation projects. The answer may be to develop better models which tie down the investor returns and link these new organisations much more tightly with existing community organisations.

## Company limited by shares

Most companies in existence in the UK are companies limited by shares. These are identical to companies limited by guarantee, except the "members" own shares in the company (and are usually known as shareholders rather than members); normally dividends are paid on the shares – so unlike a company limited by guarantee, membership brings not just the ability to vote at AGMs and other member meetings, but also a financial return.

Like companies limited by guarantee, a company limited by shares has its own legal identity and limited liability for the shareholders. The liability is limited to any sum that the shareholder has still to pay to the company for the shares; if (as is usually the case) the shares are already fully paid for, there is no further liability for the shareholder if the company goes insolvent.

Companies limited by shares are primarily used for profit-making enterprises, where the profits are distributed back to shareholders in the form of dividends. They're therefore not widely used on their own for community energy projects, where any profit-making is usually intended to be reinvested in the project (or used to take forward new projects), rather than delivering financial returns to particular individuals.

Where you might consider having a company limited by shares is as a subsidiary trading company that's wholly owned by the community organisation (e.g. a company limited by shares owned by a company limited by guarantee with charitable status). This kind of hybrid organisation is discussed in more detail on page 123.

## Community Interest Company

A Community Interest Company (CIC) is a relatively new legal form introduced in 2005. A CIC can be either a company limited by shares or a company limited by guarantee. The key difference, and one that is of particular interest for community energy, is that a CIC has additional features which go some way towards ensuring that it is run for community benefit, rather than private benefit.

The main additional features are the community interest test, the asset lock and caps on dividends or interest that can be paid. The first of these means that, when forming a CIC, the proposed directors of the CIC must submit a community interest statement to provide the CIC Regulator with some assurance that the CIC will satisfy a community interest test defined in law. The company must continue to satisfy the test for as long as it remains a CIC, and must submit a report annually to the Regulator to confirm that it is still operating within the terms of the community interest statement.

The asset lock means that a CIC must have clauses in its articles of association that restrict the transfer of any assets (including profits) so that they are primarily used for the benefit of the community. This means, in effect, that the company operates for the good of the community.

Finally, CICs can receive investment – but the amount of dividends payable to shareholders, and the amount of interest payable on performance-related loans (i.e. loans where the rate of interest is linked to the CIC's financial performance) is capped; and the cap is set by the CIC

Regulator. This means that there is a balance between encouraging investors and the benefit to the community.

CICs therefore provide all the benefits of limited companies (legal status, limited liability) but with additional features, which mean that they must operate for community benefit. For community energy projects that are looking to trade and raise finance but plough the profits made into the local community, they may sometimes be a good option. Unfortunately, however, CICs do not have any tax benefits, so it remains much more common to adopt a structure involving a charitable company limited by guarantee as the parent organisation, with a wholly owned subsidiary (which could be a CIC) carrying on the non-charitable trading activities (e.g. operating a wind farm, biomass or hydro project). Also – unlike BenComs – CICs do not currently have the benefit of any exemptions from financial services regulations relating to offer of shares to the public; so CICs are not used for the purposes of community share issues.

## Charitable status

There are three main reasons why you might want your organisation to have charitable status:

- It improves the credibility of the organisation.

- It provides a number of tax exemptions.

- It widens the range of funding sources that could be available to the organisation.

The improved credibility comes from the rules on what a charity can and cannot do, plus the increase in scrutiny

and governance that being a charity involves. This means that anyone who comes into contact with your organisation, from your members to potential grant funders, can have more faith that your organisation is being run properly and for the public good.

Provided all the money you spend as a charity is for "charitable purposes" (see page 122), your charity won't have to pay corporation tax.* Charities can also pay a reduced rate or zero rate of VAT on certain specific items (in fairness, the special VAT reliefs are relatively minor); and have the benefit of relief from land and buildings transaction tax (formerly referred to as "stamp duty"), which can be a significant saving where the organisation is buying land/buildings.

However, there are considerable downsides of charitable status as well. These include:

- More responsibility for the people involved – they become charity trustees.

- Ensuring that what you want to do fits with the eligibility requirements for being a charity.

- A significant restriction on the amount of trading an organisation can do unless a) the trading is performed in the course of carrying out a primary charitable purpose of the organisation or b) it falls below the thresholds for the "small trading exemption" or c) it is carried on by a subsidiary).

---

* Though there can be tax liabilities on the profits of a business that is not performed in carrying out a primary charitable purpose of the organisation

- Other charity rules that can restrict the entrepreneurial activities of the organisation (basically, an organisation with charitable status can only invest in new projects that do not directly further its charitable purposes if the investment can be justified as sound for the charity).

- An increase in administration, with annual returns having to be made to the charity regulator.

## Charitable purpose

In order to be a charity, your organisation needs to have what's described as a charitable purpose or purposes, and operate for the public good. There are a wide range of possible charitable purposes. Unfortunately, this range doesn't include "provide community energy" but the following charitable purposes are things that community energy projects are often seeking to do:

- Relieve poverty.

- Education.

- Health.

- Citizenship or community development.

- Protect the environment.

There are many community energy projects that have aims and purposes that can be seen as being charitable.

## Public benefit

However, having charitable purposes is not enough. The charitable purposes must also be carried out by the charity for the "public benefit". This means that you must be sure that what your organisation is doing will benefit the public in general and that any benefit to an individual is a by-product of carrying-out the purpose.

There is much room for interpretation in the area of public benefit. For example, an organisation whose purpose is "relieving poverty" is achieving public benefit, but it can only do this by relieving poverty for individuals. In this case, it can be argued that relieving poverty for individuals is a by-product of achieving the greater public benefit.

## Types of charity

Most types of organisation that we've looked at so far can register to be charities: unincorporated associations, companies limited by guarantee, and BenComs; and (in special cases i.e. where the shares are held only by a charity or charities) companies limited by shares.

It's worth noting, though, that neither CICs nor co-ops can be charities. As mentioned earlier in this chapter, the significant difference between co-ops and BenComs is that BenComs are for public benefit, whereas co-ops are for the benefit of their members.

## Charities and trading

An organisation is trading if it sells goods or services to customers. For community energy projects, trading will often take the form of selling generated electricity or heat, selling equipment (e.g. solar panels) or insulation.

A charity can trade as long as it directly furthers one or more of the "primary charitable purposes" of the charity; it is not enough to show that the net profits will be used to further primary charitable purposes – the actual trading activity has to fall within that category. In my experience this excludes most trading that community energy organisations might want to perform – which, in turn, means that most community energy organisations that trade won't be eligible to be charities.

The way around this for many is to have a hybrid organisation with a parent organisation that is entirely charitable in its activities and is therefore eligible to have charitable status, but which owns a trading subsidiary. See right for more information on hybrids.

### Charitable Incorporated Organisations

One of the downsides of having a company limited by guarantee that is a registered charity is that the organisation needs to be registered with and report to two different organisations – Companies House and the charity regulator – and also needs to comply with both company law and charity law. The Charitable Incorporated Organisation (CIO) is was introduced in 2012. It provides the benefits of incorporation (a CIO is a legal entity and provides limited liability for its members) but needs to be registered only with the charity regulator and is regulated under charity law. The legislation relating to

CIOs is much more straightforward than company law, and CIO constitutions also tend to be simpler than articles of association.

If you're thinking from the outset that being a company limited by guarantee with charitable status is the right legal form for your organisation, then you might well find that being a CIO achieves what you want a bit more easily. It's also possible to convert to a CIO, so if you currently run an unincorporated association or a company limited by guarantee and you were thinking of becoming a charity, becoming a CIO could be the best approach in the longer term.

## Hybrid structures

I've mentioned the concept of hybrid structures a few times in this chapter. This is where you have more than one legal form. Each of these is connected together to achieve the outcome you want for your organisation.

In the context of community energy, hybrid structures come about because of various tensions between the different aspects of what you want to do. Principally, you often want to have an organisation that can trade and produce an income stream, but can also provide public benefit in the form of a charity. This is not something that currently any one of the individual legal structures available in the UK can support (assuming the trading activities are not themselves directly furthering a charitable purpose – see above).

There are two hybrid structures that are most often used to overcome this:

Part 3

## Trading subsidiaries

The first is to have a company limited by guarantee that has charitable status and has a subsidiary company which does the trading.

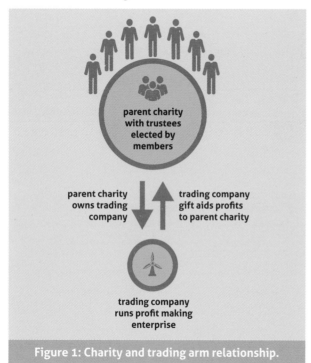

**parent charity with trustees elected by members**

**parent charity owns trading company**

**trading company gift aids profits to parent charity**

**trading company runs profit making enterprise**

**Figure 1: Charity and trading arm relationship.**

With this structure, the activities of the charity and the trading arm of the operation are kept separate. The parent charity goes about achieving its charitable aims, has a membership and a board of charity trustees.

The trading subsidiary will also be a company (for example, it could be a CIC) and it will gift-aid its profits each year to the charity (which would also mean that most – potentially all – of its profits would be free from corporation tax). The subsidiary has only one member or shareholder, which is the parent charity.

This structure allows the subsidiary the freedom to trade and be entrepreneurial, and also separates the risks that this will involve from the parent charity; but it also ties the subsidiary in with the parent charity through the parent charity's holding of shares (or its status as sole member of the subsidiary).

Extending this model, there could actually be a number of trading subsidiaries, each doing something different, but each gift-aiding its profits to the parent charity every year.

The main downside of this approach is simply the increase in the amount of time and money it takes to manage. Both companies will need annual accounts prepared, for example.

## BenComs and community benefit

The other hybrid is a bit less straightforward, but achieves a similar goal. If you want to raise money for a generation project from the public, then you may well want to go down the route of having a BenCom. However, the structure above won't work because BenComs can't be wholly owned by other companies.

The solution to this is to set up your BenCom and in its rules state that you want the "parent" charity to be the beneficiary of the activities of the BenCom. This is obviously a looser relationship than the first approach above because the charity has no direct control over the activities of the BenCom, which is controlled by its own members; and there is the risk that the members of the

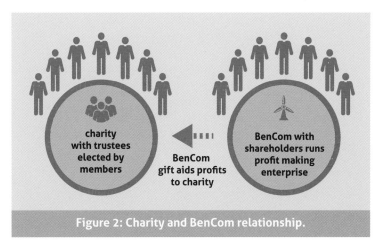

**Figure 2: Charity and BenCom relationship.**

BenCom might at some point in the future delete the provisions referring to the charity as a beneficiary. It is likely that there will shortly be moves to develop an improved version of this model.

## Geographic differences

Company law is largely uniform throughout the UK. For example, the rules governing what a company limited by guarantee or a community interest company is, and how it operates, are the same in England and Wales as they are in Scotland or Northern Ireland. From a practical point of view, the main difference is that companies in England and Wales are registered at Companies House in London or Cardiff, companies in Scotland are registered at Companies House in Edinburgh and companies in Northern Ireland are registered in Belfast.

The same is true of registered societies: the rules are the same regardless of which part of the UK you're in. If anything, for societies it's more straightforward because they all need to be registered with the FCA in London.

It's a slightly different story when it comes to charities though. While the broad approach to charity law and charities is the same, England and Wales, Scotland and Northern Ireland all have slightly different rules about how charities are defined and how they operate. For example, Charitable Incorporated Organisations were introduced in England and Wales in 2013. Meanwhile, in Scotland, Scottish Charitable Incorporated Organisations were introduced in 2011 and they don't yet exist at all in Northern Ireland (although the necessary legislation has been in place since 2008).

There are also different regulators for charities in the different parts of the UK. In England and Wales, it's the Charity Commission, in Scotland it's the Office of the Scottish Charity Regulator, while in Northern Ireland it's the Charity Commission for Northern Ireland. For the most part this shouldn't be a problem, but if you're thinking your organisation should be a charity, make sure you're reading the right information for your region.

## Legal advice

This chapter contains general information about legal matters that are relevant to community energy projects. It's important that for your own project you seek advice from a reputable firm of lawyers, ideally one that has experience with community energy and renewables.

Legal advice generally doesn't come cheap but it can save a deal of heartache (and money) further down the line.

Chapter 14

# FINANCIAL MODELS

Many community energy generation projects are set up simply to save money spent on heating or lighting a local facility. For others, though, the aim is to create a means of generating an income that can be used for other community initiatives. Often these other projects will themselves be related to energy and carbon use, but just as frequently they can have a wider scope than that. In fact, it's not unusual for there to be no specific aims at the outset for money generated, but a more general aim that it should be for the "good of the community" and for decisions to be made on a year-by-year basis.

There are also projects that aren't aiming to make money at all but are simply looking to reduce the carbon impacts of heating the village hall for example. Either way, it's important that you understand how the finances for community energy generation projects work at the outset. The aim of this chapter is therefore to look at the finances for projects that look to generate either electricity or heat.

To do this, I'm going to look at three different examples of increasing complexity:

- Firstly a solar PV project on a community hall.

- Secondly a biomass boiler in a local sports club.

- And finally, a small hydro scheme.

These projects are all fictitious but the figures used are based on real projects from around the country. Having said that, it's also fair to say that the examples here are deliberately straightforward and are looking only at the cash flow. I've made no attempt to look at profit and loss, for example, because the profits or losses your organisation might make will depend on many other aspects of what your organisation is doing.

If you're planning a project of any complexity, it's important you get professional advice, including business planning.

## Capital, FITs and RHI

All three of the projects I've used benefit from the support mechanisms currently available in the UK – FITs for the PV and hydro schemes, and the RHI for the biomass boiler. One of the features of these support mechanisms is that they pay the owners of the project for every unit of renewable electricity or heat produced. This means that you actually need to be generating before you receive anything and contrasts with the more traditional approach of having grants that help with upfront installation costs for projects.

A consequence of FITs and the RHI working this way is that the financial models for renewable energy generation projects are all based on the capital investment and paying back this capital from the income generated by the scheme. Often one of the main challenges for community energy generation projects is where to find that capital and how to pay it back while retaining money that can benefit the community.

(In fact one of the criticisms that's often legitimately aimed at FITs and the RHI is that they particularly benefit people and organisations that already have money to invest in projects. For domestic schemes this means that they operate as a kind of "middle-class subsidy", while for larger non-domestic schemes they provide opportunities for banks or other investors to reap much of the financial benefit. However, they have resulted in a significant increase in the number of renewable energy installations so they're probably not all bad.)

# Example 1: Solar PV

For the solar PV example, we're going to look at an array of panels capable of generating a maximum of 50kW sited on the flat roof of a community building that is in a town somewhere in the west of Scotland, near Glasgow. To keep things simple, in this example the building is owned and run by the local community via a company limited by guarantee that retains any profits it makes

and invests them for future projects. It has a significant heating and lighting costs each year and is looking to reduce these costs by installing the panels.

A number of local companies have been asked for quotes for the supply and installation of the panels, and each of the companies has been asked for reference sites for other work they've previously done nearby. The quote from the company that had the best combination of price and references was for £55,000 and the community group have decided that this is the one they want to go ahead with.

Because the community company that owns and runs the building has retained some profits over the years, it's in the very fortunate position of being able to afford this kind of investment without needing a loan or any other type of investment.

So we know that the capital cost of the installation is going to be £55,000, but how much money is it going to save or make the community group each year? There are three parts to the answer to this question:

- Firstly how much is the scheme going to earn from electricity exported to the grid?

- Secondly how much is the scheme going to save on electricity that the building no longer needs to buy from the grid by using the electricity straight from the panels?

- And thirdly how much is the scheme going to make from the Feed-in Tariff?

## Estimating generation

Before answering any of these questions, we need to estimate how much electricity the panels are going to generate over a year. To calculate this from first principles is horribly complex and incorporates a large number of factors, including the orientation and inclination of the panels, the geographic location, what the weather's typically like in the local area, the ambient temperature and other features of the installation itself. Thankfully there's a relatively easy way of doing it using a simple formula published in the MCS Guide to the Installation of Photovoltaic Systems,[1] and MCS-accredited installers will include this calculation in any quote so you won't need to do it yourself. However, I'm going to go through it here so that you know what's involved:

Total annual generation
= size of the scheme x Kk x shade factor

We know the size of the scheme is 50kW, but what are "Kk" and the "shade factor"?

We'll start with the shade factor. This is a measure of the amount of shade cast on the panels by other nearby objects like trees, satellite dishes and other buildings. Even a small amount of shading can have a large effect on the performance of the panels. The MCS guide mentioned above includes a diagram, which allows a reasonable estimate to be made of any shading.

The maximum the shade factor can be is 1, which means that there's no shading on the site at all at any time of day or at any time of the year. In our example, we're going to assume there are no trees or other buildings near the community centre so the shading factor is at the maximum of 1.

Kk is an estimate of the other main factors that can affect the performance of the panels. In order to know what it is, you need to know the postcode the panels are in, the orientation of the panels (i.e. how close to south are they facing?) and the pitch of the panels. Once we know all of these we can look up some tables in the MCS guide and find the correct value for Kk.

Our panels are in Scotland, near Glasgow in the G67 postcode. The tables in the MCS guide divide the country into various zones, depending on how much sunlight each of the zones receives. G67 is in zone 14, which is the West Scotland zone.

The roof of our building is measured to be at an angle of 40 degrees from horizontal, which is close to the best possible of 45 degrees. Even better, our roof faces almost directly due south – just 10 degrees to the west. With this information, we can look up the Kk value for our installation in the table for zone 14, which gives us a figure of 832.

We can now plug all these numbers into our formula to get our estimate of annual generation:

Total annual generation = 50 x 832 x 1

= 41,600kWh

Our estimate is therefore that our panels will produce 41,600 kilowatt hours of electricity per year. Having got this figure, we can move on and look at what that means in terms of income.

## 1. Feed-in Tariff

We're going to sign up to receive FITs, and the current FIT value is 4.59p[*] for every unit or kilowatt hour of electricity produced by PV installations of this size. It doesn't matter whether these units are exported to the grid or used in the building itself, we still get 4.59p for every unit.

Total annual income from FITs = 41,600 x 0.0459

$$= £1,909.44$$

## 2. Exported electricity

Now we need to work out how much of the electricity produced is used on site and how much is exported from the building. Our building uses a lot of electricity annually. As well as using some for heating and lighting, it also has air conditioning units in the offices, and fridges and freezers for storing food. We therefore estimate that approximately 70% of the electricity generated will be used on site.

If 70% is used on site, this means that 30% of the electricity will be exported and the FIT export tariff is currently 4.85p.

Total annual income from exported electricity

$$= 41,600 \text{ x } 30\% \text{ x } 0.0485$$

$$= £605.28$$

[*] Because of degression and other changes, this figure is likely to be incorrect at the time of publication. Go to the Ofgem website for the current figure.[2]

## 3. Electricity used on site

The other 70% is used on site. Currently we pay 13.5p per unit for electricity we use, so we'll be making a saving of 13.5p for every unit we use that's generated on site.

Total annual saving from electricity used

$$= 41,600 \text{ x } 70\% \text{ x } 0.135$$

$$= £3,931.20$$

### Total

To get the total financial benefit per year, we just need to add these three figures together:

Total = FIT income + Export income + Annual saving

$$= £1,909.44 + £605.28 + £3,931.20$$

$$= £6,445.92$$

This is the amount for the first year. FITs and electricity prices are all subject to inflation so will rise from year to year. However, the performance of our panels will reduce slightly each year as well. Table 1 opposite shows what the cash flow will look like if we estimate that the inflation rate will be 3% over the next 20 years and the performance of our panels will reduce by 1% each year.

You can see from this that our panels will pay back their cost in the eighth year of operation, and after that will start making a profit for our community centre.

| Year | Total generation (kWh) | Total FIT income | Total export tariff | Total saving | Total income | Total income to date |
|------|------------------------|------------------|---------------------|--------------|--------------|----------------------|
| 1 | 41,600 | £1,909.44 | £605.28 | £3,931.20 | £6,445.92 | £6,445.92 |
| 2 | 41,184 | £1,947.06 | £617.20 | £4,008.64 | £6,572.90 | £13,018.82 |
| 3 | 40,772 | £1,985.41 | £629.36 | £4,087.61 | £6,702.39 | £19,721.22 |
| 4 | 40,364 | £2,024.53 | £641.76 | £4,168.14 | £6,834.43 | £26,555.64 |
| 5 | 39,960 | £2,064.41 | £654.40 | £4,250.25 | £6,969.07 | £33,524.71 |
| 6 | 39,561 | £2,105.08 | £667.30 | £4,333.98 | £7,106.36 | £40,631.07 |
| 7 | 39,165 | £2,146.55 | £680.44 | £4,419.36 | £7,246.35 | £47,877.42 |
| 8 | 38,773 | £2,188.83 | £693.85 | £4,506.42 | £7,389.11 | £55,266.52 |
| 9 | 38,386 | £2,231.95 | £707.52 | £4,595.20 | £7,534.67 | £62,801.19 |
| 10 | 38,002 | £2,275.92 | £721.45 | £4,685.73 | £7,683.10 | £70,484.30 |
| 11 | 37,622 | £2,320.76 | £735.67 | £4,778.04 | £7,834.46 | £78,318.76 |
| 12 | 37,246 | £2,366.48 | £750.16 | £4,872.16 | £7,988.80 | £86,307.56 |
| 13 | 36,873 | £2,413.10 | £764.94 | £4,968.14 | £8,146.18 | £94,453.74 |
| 14 | 36,504 | £2,460.64 | £780.01 | £5,066.02 | £8,306.66 | £102,760.40 |
| 15 | 36,139 | £2,509.11 | £795.37 | £5,165.82 | £8,470.30 | £111,230.70 |
| 16 | 35,778 | £2,558.54 | £811.04 | £5,267.58 | £8,637.16 | £119,867.86 |
| 17 | 35,420 | £2,608.94 | £827.02 | £5,371.35 | £8,807.32 | £128,675.18 |
| 18 | 35,066 | £2,660.34 | £843.31 | £5,477.17 | £8,980.82 | £137,656.00 |
| 19 | 34,715 | £2,712.75 | £859.92 | £5,585.07 | £9,157.74 | £146,813.74 |
| 20 | 34368 | £2,766.19 | £876.86 | £5,695.10 | £9,338.15 | £156,151.89 |

Table 1: Cash flow for solar PV over 20 years.

Finally, if we're going to be very thorough, we should include a discount factor in our calculations and look at the net present value (NPV). Describing and discussing discount factors is well outside the scope of this book, but essentially it's a way of estimating how much money we expect to receive in the future is worth today.

| Year | Total generation (kWh) | Total FIT income | Total export tariff | Total saving | Total income | Total income to date | NPV income | NPV to date |
|---|---|---|---|---|---|---|---|---|
| 1 | 41,600 | £1,909.44 | £605.28 | £3,931.20 | £6,445.92 | £6,445.92 | £6,445.92 | £6,445.92 |
| 2 | 41,184 | £1,947.06 | £617.20 | £4,008.64 | £6,572.90 | £13,018.82 | £5,915.61 | £12,361.53 |
| 3 | 40,772 | £1,985.41 | £629.36 | £4,087.61 | £6,702.39 | £19,721.22 | £5,428.94 | £17,790.47 |
| 4 | 40,364 | £2,024.53 | £641.76 | £4,168.14 | £6,834.43 | £26,555.64 | £4,982.30 | £22,772.77 |
| 5 | 39,960 | £2,064.41 | £654.40 | £4,250.25 | £6,969.07 | £33,524.71 | £4,572.40 | £27,345.17 |
| 6 | 39,561 | £2,105.08 | £667.30 | £4,333.98 | £7,106.36 | £40,631.07 | £4,196.23 | £31,541.41 |
| 7 | 39,165 | £2,146.55 | £680.44 | £4,419.36 | £7,246.35 | £47,877.42 | £3,851.01 | £35,392.41 |
| 8 | 38,773 | £2,188.83 | £693.85 | £4,506.42 | £7,389.11 | £55,266.52 | £3,534.19 | £38,926.60 |
| 9 | 38,386 | £2,231.95 | £707.52 | £4,595.20 | £7,534.67 | £62,801.19 | £3,243.43 | £42,170.03 |
| 10 | 38,002 | £2,275.92 | £721.45 | £4,685.73 | £7,683.10 | £70,484.30 | £2,976.59 | £45,146.62 |
| 11 | 37,622 | £2,320.76 | £735.67 | £4,778.04 | £7,834.46 | £78,318.76 | £2,731.71 | £47,878.33 |
| 12 | 37,246 | £2,366.48 | £750.16 | £4,872.16 | £7,988.80 | £86,307.56 | £2,506.97 | £50,385.30 |
| 13 | 36,873 | £2,413.10 | £764.94 | £4,968.14 | £8,146.18 | £94,453.74 | £2,300.72 | £52,686.02 |
| 14 | 36,504 | £2,460.64 | £780.01 | £5,066.02 | £8,306.66 | £102,760.40 | £2,111.44 | £54,797.46 |
| 15 | 36,139 | £2,509.11 | £795.37 | £5,165.82 | £8,470.30 | £111,230.70 | £1,937.73 | £56,735.19 |
| 16 | 35,778 | £2,558.54 | £811.04 | £5,267.58 | £8,637.16 | £119,867.86 | £1,778.32 | £58,513.51 |
| 17 | 35,420 | £2,608.94 | £827.02 | £5,371.35 | £8,807.32 | £128,675.18 | £1,632.01 | £60,145.52 |
| 18 | 35,066 | £2,660.34 | £843.31 | £5,477.17 | £8,980.82 | £137,656.00 | £1,497.75 | £61,643.27 |
| 19 | 34,715 | £2,712.75 | £859.92 | £5,585.07 | £9,157.74 | £146,813.74 | £1,374.53 | £63,017.80 |
| 20 | 34,368 | £2,766.19 | £876.86 | £5,695.10 | £9,338.15 | £156,151.89 | £1,261.45 | £64,279.24 |

Table 2: Cash flow including NPV calculations for solar PV over 20 years.

Table 2 shows what this looks like if we use an annual discount factor of 10%. Again, this shows that we'll be making money on our panels but significantly less – the NPV break-even point won't be until the fifteenth year.

*Estimates and reality*

I haven't counted the instances but I'm aware that I've used the word "estimate" a lot in describing this example. The Kk value is an estimate so I've estimated the annual output from the panels; I've estimated how much of the electricity will be used on site and I've estimated what the annual inflation rate will be and what the reduction in performance of the panels will be. It's likely that every one of these estimates will be wrong in some way and to some degree. If you're making this kind of financial projection, it's well worth tweaking some of the values to see how much the figures can vary.

| Year | Total generation (kWh) | Total FIT income | Total export tariff | Total saving | Total income | Total income to date |
|---|---|---|---|---|---|---|
| 1 | 40,000 | £1,836.00 | £1,164.00 | £2,160.00 | £5,160.00 | £5,160.00 |
| 2 | 39,200 | £1,853.26 | £1,174.94 | £2,180.30 | £5,208.50 | £10,368.50 |
| 3 | 38,416 | £1,870.68 | £1,185.99 | £2,200.80 | £5,257.46 | £15,625.97 |
| 4 | 37,647 | £1,888.26 | £1,197.13 | £2,221.49 | £5,306.88 | £20,932.85 |
| 5 | 36,894 | £1,906.01 | £1,208.39 | £2,242.37 | £5,356.77 | £26,289.62 |
| 6 | 36,156 | £1,923.93 | £1,219.75 | £2,263.45 | £5,407.12 | £31,696.74 |
| 7 | 35,433 | £1,942.01 | £1,231.21 | £2,284.72 | £5,457.95 | £37,154.69 |
| 8 | 34,725 | £1,960.27 | £1,242.79 | £2,306.20 | £5,509.25 | £42,663.95 |
| 9 | 34,030 | £1,978.70 | £1,254.47 | £2,327.88 | £5,561.04 | £48,224.99 |
| 10 | 33,349 | £1,997.30 | £1,266.26 | £2,349.76 | £5,613.31 | £53,838.30 |
| 11 | 32,682 | £2,016.07 | £1,278.16 | £2,371.85 | £5,666.08 | £59,504.38 |
| 12 | 32,029 | £2,035.02 | £1,290.18 | £2,394.14 | £5,719.34 | £65,223.72 |
| 13 | 31,388 | £2,054.15 | £1,302.30 | £2,416.65 | £5,773.10 | £70,996.83 |
| 14 | 30,760 | £2,073.46 | £1,314.55 | £2,439.36 | £5,827.37 | £76,824.20 |
| 15 | 30,145 | £2,092.95 | £1,326.90 | £2,462.29 | £5,882.15 | £82,706.34 |
| 16 | 29,542 | £2,112.62 | £1,339.38 | £2,485.44 | £5,937.44 | £88,643.78 |
| 17 | 28,951 | £2,132.48 | £1,351.97 | £2,508.80 | £5,993.25 | £94,637.04 |
| 18 | 28,372 | £2,152.53 | £1,364.67 | £2,532.39 | £6,049.59 | £100,686.62 |
| 19 | 27,805 | £2,172.76 | £1,377.50 | £2,556.19 | £6,106.45 | £106,793.08 |
| 20 | 27,249 | £2,193.19 | £1,390.45 | £2,580.22 | £6,163.85 | £112,956.93 |

Table 3: Solar PV with different values for Kk, on-site usage and annual reduction in performance.

Table 3 on the preceding shows what happens if, for example, a more realistic Kk value is actually 800 because we're in a particularly cloudy area and in fact only 40% of the electricity produced can be used on site, and our annual reduction in performance is 2% rather than 1%.

# Example 2: A biomass heating system

Our second example is for a biomass heating system in a local sports and social club that is owned and run by the local community. Up until now, the club has been heated entirely by oil, but the price of the oil has increased significantly over the years and is costing the club a considerable amount. On top of that, the boiler was installed well over 20 years ago and needs increasing amounts of maintenance each year.

The club have looked at the options available and concluded that installing a biomass boiler fuelled by woodchip is the right way to go. It will require no changes to the pipework and radiators in the club itself but will need a new building to store the woodchip and also space for a self-contained energy cabin, which will contain the boiler itself, and a heat storage tank. Luckily the club has space at the back of the building for both of these and has applied for and been granted planning permission for the development.

The club employed a consultant who specialises in commercial-scale biomass boilers and they have produced a tender document. This has gone out to the installers in the area and a number of tenders have come

back. The best of these gives a total capital cost for all the work, including building the fuel store, of £140,000.

Unfortunately, the club doesn't have this kind of money in the bank, though they do have enough to be able to cover around 30% of the cost – about £42,000. The club have applied for a bank loan and been offered the remaining 70% of the cost, or £98,000 as a 10-year loan with a fixed interest rate of 7.5%.

## Calculating the income or saving

To calculate the annual income or saving the club's going to make from the boiler, we need to know three things:

1. How much were they spending on running the oil boiler annually?

2. How much will they spend on running the new biomass boiler?

3. How much will they receive in RHI payments?

### 1. Current running costs

Working out the existing running costs is very straightforward. The price of heating oil has changed a lot over the past few years, going both up and more recently down, but looking through old fuel bills, the club estimate they have been spending on average £30,250 on oil in each of the past three years. They have also spent around £1,000 each year on maintenance for the boiler.

So currently the annual running costs for the existing boiler are around £31,250.

*2. New running costs*

Working out the running costs for the new boiler is far less straightforward. The first thing we need to do is work out how much the annual supply of woodchip will be and in order to do that we need to work out how much heat the boiler needs to supply to the building.

Looking at the oil bills for the past few years, the club calculates that it uses 55,000 litres of oil per year. The total amount of heat produced by burning a litre of heating oil is 10.311kWh. This allows us to do a simple calculation:

Total annual heat value

= 55,000 x 10.311

= 567,105kWh

At first glance, this might appear to be the amount of heat that's supplied to the building, but bear in mind that the oil boiler won't be 100% efficient and is actually quite old so is probably closer to 80% efficient – some of the oil used won't be properly burnt and some of the heat will disappear out of the flue. Therefore:

Total annual heat supplied to the building

= 567,105 x 80%

= 453,684kWh

The new boiler will need to be supplying the same amount of heat so we now need to do the same calculation backwards to work out the amount of wood chip needed.

We expect the new boiler to be more efficient – around 90% rather than 80%. So the total amount of energy in the supplied woodchip will need to be:

Total annual heat value

= 453,684 ÷ 90%

= 515,550 kWh

The amount of heat produced by burning a tonne of good, properly dried woodchip is 3,421kWh so the total amount of woodchip we will need is:

Total annual woodchip

= 515,550 ÷ 3,421

= 151 tonnes

Like oil, woodchip costs can vary, but at the moment a tonne of woodchip costs around £120 per tonne. Therefore:

Total annual cost of woodchip

= 151 x 120

= £18,120

On top of that we have an annual maintenance charge. Looking after biomass boilers takes a bit more effort than oil or gas boilers and the club have been quoted £4,000 per year.

Total annual running costs

$$= £18,120 + £4,000$$

$$= £22,120$$

### 3. RHI income

The final part is to calculate the RHI income from running the boiler. You'll know from Chapter 9 that for commercial boilers of this scale, the RHI values are split into two tiers. The tier 1 value of 3.76p per kWh is for heat supplied up to 15% of the boiler's total annual capacity and the tier 2 value of 1.0p per kWh is for everything above that.*

The boiler has a rated capacity of 199kW so the theoretical annual capacity is this value multiplied by the number of hours in a year i.e. if the boiler was running flat-out continually for a year:

Total annual capacity

$$= 199 \text{ x } 365 \text{ x } 24$$

$$= 1,743,240\text{kWh}$$

So the tier 1 cut-off would be 15% of that value:

Tier 1 cut-off

$$= 1,743,240 \text{ x } 15\%$$

$$= 261,486\text{kWh}$$

Our annual heat requirement for the building we've calculated at 453,684kWh.

---

* Because of degression and other changes, this figure is likely to be incorrect at the time of publication

Annual FIT payments

$$= \text{tier 1 cut-off value x tier 1 rate}$$
$$+ \text{heat used above tier 1 x tier 2 rate}$$

$$= 261,486 \text{ x } 0.0376 + 192,198 \text{ x } 0.010$$

$$= 9,831.87 + 1,921.98$$

$$= £11,753.85$$

### Total

We can now do the calculation to work out what our annual income or saving is on running costs.

$$\text{Total} = \text{amount saved on oil boiler running costs}$$
$$- \text{amount new boiler costs to run}$$
$$+ \text{RHI payments}$$

$$= 31,250 - 22,120 + 11,753.85$$

$$= £20,883.85$$

### Bank loan

So on annual running costs, it looks like the club will be better off by over £20,000 per year, but this doesn't take into account how they've paid for the new boiler. If you remember, at the outset I said they'd secured a loan for 70% of the cost, which equates to £98,000. This loan is due to be paid back over 10 years and has an interest rate of 7.5%. This means that each year for the first 10 years they'll be paying back £9,800, and in the first year the interest payment will be £7,350, gradually reducing until by the tenth year it will be just £735.

| Year | Outstanding loan | Loan repayment | Interest | Total loan costs | Total income | Net income | Total income to date |
|------|------------------|----------------|----------|------------------|--------------|------------|----------------------|
| 1 | £98,000 | £9,800 | £7,350 | £17,150 | £20,884 | £3,734 | £3,734 |
| 2 | £88,200 | £9,800 | £6,615 | £16,415 | £21,510 | £5,095 | £8,829 |
| 3 | £78,400 | £9,800 | £5,880 | £15,680 | £22,156 | £6,476 | £15,305 |
| 4 | £68,600 | £9,800 | £5,145 | £14,945 | £22,820 | £7,875 | £23,180 |
| 5 | £58,800 | £9,800 | £4,410 | £14,210 | £23,505 | £9,295 | £32,475 |
| 6 | £49,000 | £9,800 | £3,675 | £13,475 | £24,210 | £10,735 | £43,210 |
| 7 | £39,200 | £9,800 | £2,940 | £12,740 | £24,936 | £12,196 | £55,407 |
| 8 | £29,400 | £9,800 | £2,205 | £12,005 | £25,685 | £13,680 | £69,086 |
| 9 | £19,600 | £9,800 | £1,470 | £11,270 | £26,455 | £15,185 | £84,271 |
| 10 | £9,800 | £9,800 | £735 | £10,535 | £27,249 | £16,714 | £100,985 |
| 11 | £0 | £0 | £0 | £0 | £28,066 | £28,066 | £129,051 |
| 12 | £0 | £0 | £0 | £0 | £28,908 | £28,908 | £157,959 |
| 13 | £0 | £0 | £0 | £0 | £29,775 | £29,775 | £187,735 |
| 14 | £0 | £0 | £0 | £0 | £30,669 | £30,669 | £218,403 |
| 15 | £0 | £0 | £0 | £0 | £31,589 | £31,589 | £249,992 |
| 16 | £0 | £0 | £0 | £0 | £32,536 | £32,536 | £282,528 |
| 17 | £0 | £0 | £0 | £0 | £33,512 | £33,512 | £316,041 |
| 18 | £0 | £0 | £0 | £0 | £34,518 | £34,518 | £350,559 |
| 19 | £0 | £0 | £0 | £0 | £35,553 | £35,553 | £386,112 |
| 20 | £0 | £0 | £0 | £0 | £36,620 | £36,620 | £422,732 |

Table 4: Biomass cash flow.

Net income/saving in first year

$$= £20,883.85 - £9,800 - £7,350$$

$$= £3,733.85$$

*First 20 years*

Like electricity and FIT payments, oil, woodchip, boiler maintenance and RHI are all subject to inflation. If we estimate that the inflation rate will be 3% then our cash flow model for the first 20 years of operation looks like table 4.

The break-even point for the club looks like it will be in year 7, but like the solar photovoltaic example, there have been quite a few estimates and assumptions made in these calculations so it's a good idea to change some of the figures a bit and see what effect that has on the cash flow.

# Example 3: A small hydro scheme

The final example we're going to use is for a small hydro scheme. Again, this is a fictitious example, but based on a number of real projects around the country. The example we're going to use is a 50kW scheme that is going to extract water from a stream, and run it in a pipe down the hill next to the stream to a shed that contains the turbine. Having passed through the turbine, the water then re-enters the stream.

The first thing to note about projects like these is that they really need to be considered in two or possibly three stages. There's the very first "feasibility" stage where the idea of having a project is looked at to check that it will actually be feasible and there are no obvious, serious problems with it that would mean it's not worth taking any further.

The second stage is where the detailed planning for the project's done. Projects like this need serious design work: planning permission will be required; an agreement will have to be reached with the local landowner (or possibly landowners); and environmental agencies

will need information about what effect the scheme will have on the local environment. The only way to do all of this is to employ professionals, which means that this stage can cost a significant amount of money.

The problem has always been that money spent at this stage is "at risk" because if you can't get planning permission, reach an agreement with a landowner or satisfy the environment agency, the project can't go ahead and you'll have nothing to show for the money you've spent. This is fine for organisations or people who are relatively wealthy and can afford to take these kinds of risks, but for start-up community organisations, trying to find money for this stage was a serious barrier to many projects getting beyond the "nice idea" stage.

More recently, however, some government schemes have been set up to help alleviate this problem. In England there are the Rural Community Energy Fund (RCEF) and the Urban Community Energy Fund (UCEF) while in Scotland there's the Community and Renewable Energy Scheme (CARES) (see Chapter 4). Although different in the details, all of these schemes work the same way in principle: they provide a loan to community groups for this planning stage of projects. If a project is successful and goes ahead, then the loan must be repaid; if a problem's encountered that stops the project from going, ahead then the loan's written off.

(Although it's less significant from the point of view of the example here, these schemes also provide grant funding – i.e. not loans – for the very initial feasibility work.)

## Some numbers

So, what are the key statistics for our example project? As mentioned previously, it's a 50kW hydro scheme. Consultants employed to work on it have estimated that it will have a capacity factor of 65%, which means that over a year, it will produce 65% of the total maximum electricity it could theoretically produce if it was working flat-out all the time. Therefore:

Total annual energy produced

= size of scheme

x number of hours in a year

x capacity factor

= 50 x 24 x 365 x 65

= 284,700 kWh

We've got planning permission, an agreement with the landowner, a grid connection agreement, and the Environment Agency are satisfied with the plans. To achieve this has cost £40,000, which we've paid for with an RCEF loan. The way RCEF and UCEF loans work is that you pay them back when you raise the rest of the finance for the project, but we pay them back with a 45% premium, which for a £40,000 loan is £18,000.

We've had quotes from installers and the total cost of installation is going to be £250,000. Adding this figure to the RCEF loan and premium means that the total capital cost for the project will be £308,000.

We've set up a BenCom (see page 118) to run the project and we're planning to raise £200,000 via a share issue to local people. We're aiming to give the shareholders an interest rate on their shares of 4%. However, we're not going to start doing this until the third year of operation. This is for two reasons – firstly so that we can start creating a reserve fund, and secondly so that we have more money available for other community projects early on.

The rest of the capital we'll raise from a bank and we've been offered a 10-year loan with an interest rate of 7.5%.

## Calculating annual income

Calculating the annual income for the scheme is fairly straightforward. We have an estimate for how much electricity the scheme should produce. All this electricity will be exported so we only have to calculate the income from the FIT and the export tariff because none of it is offsetting electricity used elsewhere. For schemes of this size, the FIT is currently 8.54p per kWh and the export tariff is 4.85p per kWh.

Annual FIT income

= 284,700 x 0.0854

= £24,313.38

Annual export tariff income

= 284,700 x 0.0485

= £13,807.95

Total annual income

= £24,313.38 + 13,807.95

= £38,121.33

However, there are also some ongoing costs in the scheme, namely annual maintenance costs of £3,500 and other ongoing costs covering the land lease, insurance, administration and council tax that amount to £5,000 per year so we need to subtract these from our income:

Net income  = £38,121.33 – £3,500 – £5,000

= £29,621.33

As with our previous two examples, the FIT payments, the maintenance and all the other costs are subject to inflation so if we expand this to look at the first 20 years, we get Table 5, overleaf.

## Loans and interest

We're expecting to have a bank loan of £108,000 to be paid back over 10 years and with an interest rate of 7.5%. This means that each year for the first 10 years of operation we'll be paying back £15,400 and in the first year paying £8,100 in interest, which will steadily drop to £810 in the tenth year.

We've said that we're aiming to pay 3% interest to our shareholders, but only from year three. This will therefore be zero in the first two years before increasing to £6,000 and staying steady at this rate.

Table 6 overleaf shows what the "investor costs" look like over the 20-year period.

## Putting it all together

If we put it all together, we end up with the cashflow in Table 7.

## Reserve

The community group decided that keeping a significant amount of money in reserve is a good thing. This is for a number of reasons:

- If there's a drought one year, the income will drop significantly. They need to able to maintain loan and interest payments if this happens.

- Some of the shareholders might want to sell their shares back to the company. The company will look to resell these, but there is the possibility that it won't be able to.

- There might be a significant problem with the scheme that will cost a lot to put right.

The group decide that generating a reserve of £30,000 is a good idea, but that this should be reviewed after the first 10 years after the bank loan has been repaid. However, they also want to start making some money available for other community projects as soon as possible, so they agree to build up this reserve over the first four years.

Table 8 shows the amount of money available for community projects once this is taken into account.

Again, the same health warning applies that these are projections with considerable estimates and assumptions, and as mentioned above a long period of dry weather can have a significant effect on a hydro project.

| Year | Maintenance cost | Other costs | Total ongoing costs | Gross income | Net income |
|------|------------------|-------------|---------------------|--------------|------------|
| 1 | £3,500 | £5,000 | £8,500 | £38,121 | £29,621 |
| 2 | £3,640 | £5,150 | £8,790 | £39,264 | £30,474 |
| 3 | £3,785 | £5,304 | £9,090 | £40,442 | £31,352 |
| 4 | £3,937 | £5,463 | £9,400 | £41,656 | £32,255 |
| 5 | £4,094 | £5,627 | £9,722 | £42,905 | £33,183 |
| 6 | £4,258 | £5,796 | £10,054 | £44,193 | £34,138 |
| 7 | £4,428 | £5,970 | £10,398 | £45,518 | £35,119 |
| 8 | £4,605 | £6,149 | £10,755 | £46,884 | £36,129 |
| 9 | £4,789 | £6,333 | £11,123 | £48,290 | £37,167 |
| 10 | £4,981 | £6,523 | £11,505 | £49,739 | £38,234 |
| 11 | £5,180 | £6,719 | £11,900 | £51,231 | £39,331 |
| 12 | £5,388 | £6,921 | £12,309 | £52,768 | £40,459 |
| 13 | £5,603 | £7,128 | £12,732 | £54,351 | £41,619 |
| 14 | £5,827 | £7,342 | £13,170 | £55,982 | £42,812 |
| 15 | £6,060 | £7,562 | £13,623 | £57,661 | £44,038 |
| 16 | £6,303 | £7,789 | £14,093 | £59,391 | £45,298 |
| 17 | £6,555 | £8,023 | £14,578 | £61,173 | £46,594 |
| 18 | £6,817 | £8,264 | £15,081 | £63,008 | £47,926 |
| 19 | £7,090 | £8,512 | £15,602 | £64,899 | £49,296 |
| 20 | £7,373 | £8,767 | £16,141 | £66,845 | £50,704 |

Table 5: Simple annual income for a 50kW hydro scheme.

| Year | Oustanding loan | Loan repayment | Interest | Total loan costs | Shareholder payments | Total investor costs |
|---|---|---|---|---|---|---|
| 1 | £108,000 | £10,800 | £8,100 | £18,900 | £0 | £18,900 |
| 2 | £97,200 | £108,00 | £7,290 | £18,090 | £0 | £18,090 |
| 3 | £86,400 | £10,800 | £6,480 | £17,280 | £6,000 | £23,280 |
| 4 | £75,600 | £10,800 | £5,670 | £16,470 | £6,000 | £22,470 |
| 5 | £64,800 | £10,800 | £4,860 | £15,660 | £6,000 | £21,660 |
| 6 | £54,000 | £10,800 | £4,050 | £14,850 | £6,000 | £20,850 |
| 7 | £43,200 | £10,800 | £3,240 | £14,040 | £6,000 | £20,040 |
| 8 | £32,400 | £10,800 | £2,430 | £13,230 | £6,000 | £19,230 |
| 9 | £21,600 | £10,800 | £1,620 | £12,420 | £6,000 | £18,420 |
| 10 | £10,800 | £10,800 | £810 | £11,610 | £6,000 | £17,610 |
| 11 | £0 | £0 | £0 | £0 | £6,000 | £6,000 |
| 12 | £0 | £0 | £0 | £0 | £6,000 | £6,000 |
| 13 | £0 | £0 | £0 | £0 | £6,000 | £6,000 |
| 14 | £0 | £0 | £0 | £0 | £6,000 | £6,000 |
| 15 | £0 | £0 | £0 | £0 | £6,000 | £6,000 |
| 16 | £0 | £0 | £0 | £0 | £6,000 | £6,000 |
| 17 | £0 | £0 | £0 | £0 | £6,000 | £6,000 |
| 18 | £0 | £0 | £0 | £0 | £6,000 | £6,000 |
| 19 | £0 | £0 | £0 | £0 | £6,000 | £6,000 |
| 20 | £0 | £0 | £0 | £0 | £6,000 | £6,000 |

Table 6: Investor costs for a 50kW hydro scheme.

| Year | Total investor costs | Net income after ongoing costs | Net income received | Total income to date |
|---|---|---|---|---|
| 1 | £18,900 | £29,621 | £10,721 | £10,721 |
| 2 | £18,090 | £30,475 | £12,385 | £23,106 |
| 3 | £23,280 | £31,353 | £8,073 | £31,179 |
| 4 | £22,470 | £32,256 | £9,786 | £40,965 |
| 5 | £21,660 | £33,184 | £11,524 | £52,489 |
| 6 | £20,850 | £34,138 | £13,288 | £65,777 |
| 7 | £20,040 | £35,120 | £15,080 | £80,857 |
| 8 | £19,230 | £36,129 | £16,899 | £97,756 |
| 9 | £18,420 | £37,167 | £18,747 | £116,503 |
| 10 | £17,610 | £38,234 | £20,624 | £137,128 |
| 11 | £6,000 | £39,331 | £33,331 | £170,459 |
| 12 | £6,000 | £40,460 | £34,460 | £204,919 |
| 13 | £6,000 | £41,619 | £35,619 | £240,538 |
| 14 | £6,000 | £42,812 | £36,812 | £277,350 |
| 15 | £6,000 | £44,038 | £38,038 | £315,388 |
| 16 | £6,000 | £45,299 | £39,299 | £354,687 |
| 17 | £6,000 | £46,595 | £40,595 | £395,281 |
| 18 | £6,000 | £47,927 | £41,927 | £437,208 |
| 19 | £6,000 | £49,296 | £43,296 | £480,505 |
| 20 | £6,000 | £50,704 | £44,704 | £525,209 |

Table 7: Combined cash flow for a 50kW hydro scheme.

| Year | Net income | Money kept in reserve | Money available | Total to community |
|------|-----------|----------------------|-----------------|-------------------|
| 1 | £10,721 | £10,000 | £721 | £721 |
| 2 | £12,385 | £10,000 | £2,385 | £3,106 |
| 3 | £8,073 | £5,000 | £3,073 | £6,179 |
| 4 | £9,786 | £5,000 | £4,786 | £10,965 |
| 5 | £11,524 | £0 | £11,524 | £22,489 |
| 6 | £13,288 | £0 | £13,288 | £35,777 |
| 7 | £15,080 | £0 | £15,080 | £50,857 |
| 8 | £16,899 | £0 | £16,899 | £67,756 |
| 9 | £18,747 | £0 | £18,747 | £86,503 |
| 10 | £20,624 | £0 | £20,624 | £107,128 |
| 11 | £33,331 | £0 | £33,331 | £140,459 |
| 12 | £34,460 | £0 | £34,460 | £174,919 |
| 13 | £35,619 | £0 | £35,619 | £210,538 |
| 14 | £36,812 | £0 | £36,812 | £247,350 |
| 15 | £38,038 | £0 | £38,038 | £285,388 |
| 16 | £39,299 | £0 | £39,299 | £324,686 |
| 17 | £40,595 | £0 | £40,595 | £365,281 |
| 18 | £41,927 | £0 | £41,927 | £407,208 |
| 19 | £43,296 | £0 | £43,296 | £450,505 |
| 20 | £44,704 | £0 | £44,704 | £495,209 |

Table 8: Total income for community projects.

Part 3

The financial viability of this project is very much dependent on how much money you can raise from private investors.

# Tax

I haven't mentioned anything about tax of any sort in these examples, but it's worth being aware of what the situation is.

## VAT

Income from FITs and the RHI is outside the scope of VAT and therefore no VAT needs to be declared on the FIT income. However, the export tariff is deemed to be a supply of electricity and a VAT-registered business supplying electricity needs to add VAT at 20% to the export tariff. In terms of the long-term cash flows presented here, this makes little difference because the VAT gathered will be accounted for and paid to HMRC each quarter.

VAT at 20% is also payable on the supply and installation of all the equipment discussed in this chapter. The assumption has been that all the fictitious organisations in this chapter are well over the threshold for VAT registration so are registered for VAT and can therefore claim the VAT back; but organisations not registered for VAT can find themselves paying considerably more for the equipment and its installation.

## Corporation tax

For organisations, income from FITs and the RHI is taxable so needs to be declared as part of your income in your annual accounts.

Organisations can receive capital allowances for the cost of renewable installations as for other items of capital expenditure.

## Personal taxation

Although this chapter has focused on organisations, it's worth noting that income individuals receive via FITs and the RHI is exempt from personal taxation provided the electricity or heat generated is mainly for their own use.

# Sources of finance

As mentioned earlier in this chapter, there are two points during a community energy generation project at which the community group need to access significant amounts of money. The first of these is for larger projects when considerable amounts of work need to be done planning and gaining permission for the project. The second is to cover the cost of the supply and installation of the scheme itself.

## Risk money

"Risk money", is the money that's needed when the detailed planning for a project is taking place, but before it has been granted all the permissions required to go ahead. This was traditionally a major problem for community groups, because they very rarely had money for this activity and even if they did, the money was very much "at risk", i.e. it's highly possible that at this stage the money would be spent, but an obstacle to the project going ahead would have been found.

It's to help alleviate this problem that a number of government schemes have been set up for community groups. In England the Rural Community Energy Fund (RCEF) and the Urban Community Energy Fund (UCEF) both provide loans of up to £130,000 for a variety of activities that need to take place at this stage. This money then needs to be paid back with a premium of 45% when the project reaches "financial close", i.e. when all the money for building the scheme has been raised.

In Scotland, there is the Community and Renewable Energy Scheme (CARES). This provides loans of up to £150,000 for pre-planning activities. These loans have a fixed interest of 10% attached to them.

What RCEF, UCEF and CARES have in common is that the loans are written off if the project fails. This can take much of the worry and risk out of projects for community groups. There's more information about these schemes in Chapter 4.

In Wales the Ynni'r Fro community programme provides grants of up to £30,000 for pre-planning activities. These are grants rather than loans so they don't need to be paid back even if a project is successful. At the time of writing there's no equivalent scheme in Northern Ireland.

## Capital

The largest amount of money you'll need for nearly any generation project is the capital for building whatever it is you're doing. With FITs and the RHI now providing a reliable and fairly predictable source of income, the number of options available has increased and you can take a number of different approaches, some of which have been touched on in the examples in this chapter.

Increasingly popular at the moment are share issues for BenComs. These allow local people to invest in projects and receive a modest return on that investment for the lifetime of the project. The legal structure of BenComs is described in the previous chapter. Until recently there were also tax advantages for people to invest in BenComs, although this has now changed.

Getting a loan for at least some of the capital from a bank is another route. It used to be that the Co-operative Bank was the first port of call for community energy projects, but with the problems that bank has had, it has completely withdrawn from the renewable energy market including community renewables. It's always worth talking to the big banks and also some of the smaller ones like the Charity Bank and Triodos.

In Scotland, the Renewable Energy Investment Fund (REIF) can provide loans to community renewable generation projects.

Another approach is "crowdfunding". Crowdfunding" is the term used for the practice of funding a project or venture by raising many small amounts of money from a large number of people. This is typically done on the internet.

Part 3

A number of crowdfunding companies have set-up providing crowdfunding internet sites or "crowdfunding platforms". These include Abundance Generation which provides crowdfunding for renewable energy generation for both community and straightforward commercial projects.

Grant funding is currently very unusual for community energy projects. Financial support from the government for renewable energy is primarily through FITs and the RHI, and in fact it's not possible to receive a government grant and FITs or the RHI for the same project. Grant funding from other bodies is going to depend very much on what the aims of the project are.

If you can demonstrate that your project will have a significant impact on households in fuel poverty or low-income households, then there are non-government grant bodies that might invest in your project, but you'll have to have a very good case before this will happen. This also applies to Energy Company Obligation (ECO) funding from the big energy companies as well, which can be a good source of money if your project ticks the right boxes.

Chapter 15

# SOME PRACTICALITIES

## Smaller projects

Many community energy projects are relatively small and straightforward for example – a single solar array, or a heat pump for a community building. The main motivation for this kind of project will very often be financial – saving or making money for an existing community organisation by generating your own electricity or increasing the efficiency of your heating system and receiving the FIT or RHI.

Because these types of schemes are often being added to existing buildings, there is frequently no need for a separate organisation to be created to manage and own the installation – the existing organisation that owns or runs the building can fulfil those roles.

Not so long ago, installing a renewable energy system was taking something of a step into the unknown. Things have changed rapidly, and these days it's like the installation of any other piece of building equipment. That said, you'll still be spending thousands of pounds and you should be thinking carefully and systematically about what you're doing, and you'll need to go through a procurement process to ensure that what you're fitting is right for your particular situation and that the installer you choose is capable of doing the job.

A small project can be a good starting place for a community group that wants to get into community energy. You'll be solving a specific problem: improving a community asset or improving the cash flow for a community organisation – things that people within a community find easy to understand and easy to support.

## Not-so-small projects

Any larger project like a small hydro scheme, a community wind turbine or a street-by-street retrofit programme will be significantly more complex. Renewable energy projects can frequently take a number of years and there will always be obstacles and problems that arise along the way. The amount of time and effort required by the volunteers shouldn't be underestimated, nor should the pressure that they will frequently feel: as well as investing time, they will also make a considerable emotional investment in the project.

Many veterans of both successful and unsuccessful projects describe the experience as being like a roller-coaster with fantastic highs when a milestone like planning permission is reached, but equally terrible lows when it feels like the project has reached the end of the line and the dream will have to be abandoned. My view is that there's no avoiding this aspect of the experience, although it helps to be prepared both for the level and length of commitment that might be required and for the emotional rollercoaster you're about to willingly climb aboard!

There's no recipe that guarantees a project will succeed, but there are many things you can do to make it more likely.

# Your group

The first thing you need to think about is how your group works. As mentioned above, the time and emotional investment that gets put in to community energy projects can be very high. However, it's important that no-one promises time they haven't got and no-one should feel under pressure to commit to things that they can't manage. Everyone needs to be realistic about what might be required and what they're able to provide.

## Your skills

It's also important that as a group you're aware of your strengths and weaknesses and how well the skills you've got map on to the skills that will be required. Between you, you'll need to have good organisational and communication skills, a great deal of patience and also significant amounts of determination. Among the skills that might be useful are:

- Running a small business and writing a business plan.

- Commissioning consultants.

- Knowledge of renewable energy.

- Experience in and an understanding of the planning system.

- Experience in fundraising.

- Knowledge of bookkeeping and accountancy.

- Experience in project finance.

- Dealing with the media.

- Running events and giving presentations.

- Facilitating public debates.

- Experience in setting up and running websites.

This isn't an exhaustive list and it isn't vital to have all of these skills within your group either: some of these things you might never need and some you would hire a commercial organisation to do for you. (In particular, for anything but the smallest and most straightforward projects, you're going to be hiring renewable energy consultants.) However, it's worth being aware of what skills you have in your group, where there might be gaps, how significant those gaps are, and what you plan to do about them.

## Your organisation

How you organise your group is going to depend on both the make-up of the group itself and what it is you're planning. It might be that you want to have subgroups with responsibility for different aspects of your project – one for planning, another for finance, and so on. It might be that you want to meet monthly or possibly weekly when there is a lot happening. It might be that you want to be quite formal in how you run your meetings, with an agenda and minute-taker, or it might be that you want to have a more relaxed approach. All of these things will vary from project to project and will probably vary over the lifetime of a project as well depending on the level of complexity and level of activity at the time.

You also need to consider how important it is to your group that the experience is fun and enjoyable. For some groups this won't be a significant element at all, but for many it will be. The initial conversations that become community energy projects often take place in social settings – in the pub or at the school fête – and for many groups this social aspect continues to be important and shouldn't be lost. Your group might continue to meet in the pub and this social, fun aspect of the project can be an important ingredient – people are more likely to remain involved and committed to a project if they have a good time doing it. However, like all these things, it's a question of balance. At times, in all projects, serious far-reaching decisions will have to be made and it's important that the people making the decisions are focused and understand what's being decided. At other times, being relaxed and sociable can result in highly creative solutions for apparently intractable obstacles. Some groups will have a serious business-like approach, others will prefer to be more genial and gregarious.

### Legal identity

At some point, probably fairly early on in your project you'll need to decide what legal structure (or even structures) is right for your project. This isn't as scary as it might seem and very often the right type of organisation is obvious.

It's also the case that you don't necessarily need to set up your legal identity straight away. It could be that you know what you want to be, but you don't necessarily have to create that identity until you're confident that your project has a chance of success or you need to enter into a legal agreement with another party.

The different forms of legal identity are described and discussed in more detail in Chapter 13.

# Gaining permission

In the course of a renewable energy generation project you'll usually have to consider getting planning permission for what you want to do. This prospect fills some people with dread, but in reality the process is not as bad as all that, though it does help to be organised and patient.

Not every development needs planning permission though. Smaller and less conspicuous renewable energy installations like some solar, heat pumps and wood-burning stoves fall into the category of "permitted development", meaning you don't need to submit a planning application.

For those projects that do need planning permission, this is granted (or withheld) by the local authority. Details of the process of applying for and being granted planning permission varies, depending on which part of the country you're in, but the general principles are the same, and how the process should be approached is the same as well.

Firstly and most importantly (for your own sanity, as well as the good of the project), you should seek to develop a good relationship with your local authority. Viewing the local authority as an obstacle and the council officials as obstructions to be overcome is rarely the right way to approach this issue. It's also a mistake to believe that because your project is for the greater good of your community and the wider world, it must be granted planning permission. In an absolute moral sense you might be right, but approaching the local authority with this mindset will only cause you problems. Council officials have statutory duties to protect or preserve things that are important to people, like landscape, noise levels or heritage. Even if they wanted to, these officials can't change the local planning policy to accommodate your project. It's much better to try and work with the officers, and if they identify problems with your project, understand how you can amend them in a way that makes them acceptable.

You should also be prepared to work with more than just one person in the council. Who these people are is going to vary from project to project and council to council, but among them might be:

- Planning Policy Officers who don't deal directly with planning applications, but are responsible for developing the long-term plans against which the Development Management Officers will judge your application. Planning Policy Officers may themselves be asked to comment on it as well.

- Development Management Officers who will actually deal with your planning application and advise you through the process of making the application. In simple cases they will issue the decision themselves (on behalf of the council), while in more complex or controversial cases they'll make a recommendation to the elected councillors on whether permission should be granted. Their decision will be based on the long-term plans developed by the Planning Policy Officers as well as the opinions of other officers in the council and very possibly statutory consultees as well (see the next page).

- Landscape Officers who will look at any landscape impacts your project might have and advise on any issues that arise.

- Ecology Officers who will assess any impacts your plans might have on the local ecology – the wildlife and plants that live around your project.

- Historic Buildings or Conservation Officers who will be involved if your project involves or affects any listed buildings or other heritage features and are responsible for deciding whether Listed Building Consent or Conservation Area consents should be granted.

There are others that you might have to work with as well. For example, if your project requires an access road, you might end up speaking to someone from the highways department.

Depending on the nature of the project, there could well be some statutory consultees involved as well. These are other organisations that the council is obliged to consult with on planning applications and who provide expert opinion on different aspects of the project in which they have an interest. From the long list of statutory consultees that exist you might encounter the Environment Agency, English Heritage, the Forestry Commission, Historic Scotland, Scottish Environmental Protection Agency (SEPA), the Countryside Council for Wales, and so on. Again, it's important to try and work with the officials from these organisations.

For a project of any size, the actual contents of the planning application are likely to be produced by a professional firm of consultants who you've commissioned to do the work. The amount of time and expertise required to put together a planning application, combined with all the supporting information that is required, means this isn't something that can be done in your spare time. It obviously makes sense for these consultants to have direct contact with the council officials and indeed in some cases the consultants will take the lead in this relationship. However, even in these situations it makes sense for you as the client to remain involved, ensuring you understand what's happening so that, when required, you can make informed decisions.

It's a good idea to meet with the planning officers at your local authority early on to talk through your proposal and discuss how well it fits with current planning policies and what information they'll need as part of a planning application. Provided you've worked with the council officers in preparing the application, by the time you submit it, any issues should already have been raised,

discussed and where possible addressed. You should be aware of what the chances are of permission being granted, and the outcome, when it happens, shouldn't be a surprise even if it's not what you wanted.

## Changing planning policy

Local authorities regularly update their planning policies in the form of Local Development Frameworks or Local Development Plans, or in the form of other planning policy documents that have direct relevance to renewable energy projects (e.g. defining areas of search). The public has the right to take part in the preparation of these documents. It could be that you know a project you have in mind won't be granted planning permission at the moment because the existing planning policies won't allow it. You might choose therefore to get involved in the process of updating planning policies with the aim of improving the chances of your project being successful in the future.

Getting involved in changing planning policy so that your project can go ahead is obviously a very long way round to achieve your aims and with no guarantee of success at the end of the process. When setting planning policy, the planning officers will be trying to balance many different factors, so the outcome is far from certain.

If you do want to get involved in this process, you should contact your local council and find out what's currently going on and when public consultations are going to take place. The important part is understanding what to say and when, so check the proposed dates and speak to the planning policy team and ask to be kept up to date with any changes to the timetable.

# Support for your project

Renewable energy generation projects can often be highly controversial and it's a good idea to look at how you can gain support for your project.

## Within your community

To your group it might seem obvious that your project is a good thing with both local and global benefits, but others in your community might not see it in the same way and you should be prepared to put time and effort into taking your community with you. There are many different ways of doing this and which of the approaches is best for you and your project will depend on the local circumstances.

Consulting within your community can be an important step in making a success of your project. Community consultation itself comes in many shapes and sizes.

You could develop the plans for your project in some detail, present them at public meetings within the community and then answer any questions raised. This is probably the approach most often taken and in the right circumstances there's nothing wrong with it. However, you need to be confident that people will be broadly enthusiastic about your project before you do this because you're not actually giving them the opportunity to get involved and help shape what you're doing. The risk with this is that some might not like what you're proposing. This can result in conflict with people in

"for" and "against" camps, taking increasingly entrenched positions. Ultimately, this situation can completely derail your project and in the worst cases can cause a great deal of ill feeling within communities that lasts for years.

To avoid this type of thing happening, it's important to be as open as possible to thoughts and ideas from other people, and genuinely consult with them about what you're thinking of doing in the early stages.

It's worth thinking about the network of influence of the people within your group. The chances are you'll want to talk to other people and organisations within your community, like parish councillors or the local school. There will be people outside your group who you want to discuss your plans with. These discussions should be two-way: they can help you to gauge likely interest or antipathy towards your plans and they can also help you influence key people within your community. In the same way as looking at your collective skills, it's worth looking at who you know so that you can speak to people informally about what you're planning to do.

Of course, there is a reality that for various reasons some people will always oppose your plans. However, broad-based and open consultations can help reveal common ground between groups who, on the face of it, appear to have little in common. One of the subjects that tends to pull people together is fuel bills. Framing your discussions around the cost of energy can work wonders and, by way of contrast, sadly, framing it around climate change can have the opposite effect. It's worth thinking about what themes will bring people on board before starting your consultation.

There is, though, one other risk with open, inclusive consultation and that is "death by consensus". It's unlikely that you'll ever achieve a full consensus within your community about your plans and trying to achieve consensus or something close to it can slow the whole process down so much that you lose critical momentum, or you end up with endless meetings attempting to resolve the same unresolvable issues.

## Outside your community

It's also worth thinking about who outside your community you might want to support your plans: in particular local councillors (especially those for your own ward), your local MP, and in Scotland, Wales and Northern Ireland your MSPs, AMs and MLAs respectively.

A first step would often be to arrange a meeting to let them know what you're planning and to discuss these plans with them. It's easy to be cynical about politicians, but most elected representatives are keen to support activities that will benefit their local areas. However, a bit like framing your community consultation, it can be worth framing your conversations in this context as well. There may well be parts of your project that they feel more comfortable with than others and having it presented to them in this way can make for an easier conversation. It's important not to be deceitful in any of these conversations though. Don't hide things about your project but, for example, start by presenting the local problems it will solve or the local benefits it will provide rather than the details of the project itself.

Although MPs, MSPs, and so on, are not usually in a position to directly affect your project, their visible support in local media can help add credibility and weight. They can also fulfil a very useful role in unlocking doors to people you might otherwise have trouble reaching. If a project is unable to proceed for technical or planning reasons, there won't be much they can do but if, for example, you're having problems arranging a meeting with someone, a phone call from your MP to the person in question can make things happen.

It's obviously worth making contact with your local councillors as well. If any of your councillors will be involved in making a decision on your planning application, then there are strict limits on what they can do and say about your project, but there's nothing to stop you giving them information about it and discussing with them what your plans are and what the benefits will be even if they can't comment on what they think of it to you or anyone else.

# Raising finance

One of the characteristics of energy generation projects is that they can provide a long-term income, but you still need to find the funding or capital so that you can build your scheme in the first place. Unless you're in the very unusual and very fortunate position of having an organisation that already has considerable funds at its disposal, you'll need to persuade someone else to commit the funds to back your idea. There are various possible sources for this funding and there are benefits and downsides to each of them and any potential source, but they will be making hard-headed decisions about whether they want to be involved in your project.

Part 3

The three main potential sources of finance are grant funding, share issues and loans.

## Grant funding

Grants are sometimes seen as the most desirable form of funding because it's money that doesn't have to be paid back. While that's usually true, all grant funders have strict criteria for how their money is spent and will require specific outcomes be achieved as a result of their investment – outcomes that you will have to have committed to when you applied for the grant. This means that you will need to keep the funders in touch with what you're doing in the form of regular reports and you will also need to ensure that you are achieving what you promised.

There are therefore costs attached to grants, both in terms of the time required to do the reporting and in terms of the restrictions on your freedom to run your project the way you want to, but if your project's aims align well with the requirements of the grant, it can be a great way to find the funding or at least some of the funding that you need.

There are a number of different sources for grants. Firstly independent charitable trusts provide a huge amount of grant funding within the UK. Most of the big ones have excellent websites that describe their aims, the types of grants they provide, their criteria and the application processes. You should have a good read of these before progressing with any application. There are also a number of websites that provide databases of charitable trusts including trustfunding.org.uk, and cafonline.org .

The next major source of grants is government, both national and local. The Big Lottery Fund is definitely worth looking at, but they won't fund renewable energy projects for the sake of it. Whether an application to one of the Big Lottery funds is successful or not, is going to depend, to a large extent, on what the outcomes of the project will be.

The national government is more focused on the FIT and RHI support mechanisms than providing grants and even if grants are available, you need to be careful that receiving a grant doesn't have the effect of making your project ineligible to receive FITs or RHI payments under European rules.

Your local council may also have grant schemes, some of which they manage on behalf of others. For example, if your area has a lot of quarrying, there may be funds available for improving the local environment that come through the taxes levied on quarrying.

It might also be possible to receive money from commercial organisations. Most large corporations these days run Corporate Social Responsibility programmes, often referred to as CSR. Through these, charitable ventures can often access significant funding or professional support.

The UK government's Energy Company Obligation (ECO) scheme places legal obligations on larger energy suppliers to deliver energy-efficiency measures to domestic premises. It focuses on insulation and heating measures and aims to support consumer groups. If your project fits with the ECO criteria, it may well be possible for it to receive funding under this scheme. All the large energy companies have teams dedicated to the delivery of the ECO requirements.

## Share issues

Issuing shares in your scheme is an increasingly popular way of raising the capital needed to make projects happen. As described in Chapter 3, this is an approach that was pioneered in the UK by the co-operative Baywind and has now been used by a large number of schemes, usually organised as co-operatives or BenComs.

One of the advantages of this approach is that it allows people who live in your community to have a direct financial investment in your project. Having a group of shareholders can be a really valuable way of increasing involvement in your project and of growing local support: they quite literally have an investment in its success. One of the downsides is that people who can invest a greater amount benefit more than those in your community who can't afford to invest so much or anything at all.

If you take this approach, it's important that the point of your project doesn't become the return investors can receive. You should retain a focus on the social and environmental value of the project and an appeal to the "social", "ethical" or "community" oriented aspect of the investment.

Raising money from a share issue is a heavily regulated business so you'll need professional advice. The regulations are generally designed to make sure that anyone investing has been told about all of the risks involved in the project and hasn't been given a misleading picture about its chances of success or the likelihood of rewards.

However, the number of projects that are now funded in this way, demonstrates that it's a viable approach for a community project.

## Loans

The final method of raising money is to borrow it. If you want to borrow money for your project, you will need to convince whoever is lending you the money that you will be able to repay them, and you will need to provide security to cover the full value of the loan in case the repayment plan fails.

It's highly unlikely that you will manage to raise all the capital required for your project as a loan, so loans will almost always be combined with some other source of finance, a share issue for example.

The obvious source for a loan is a bank and banks are becoming increasingly aware of the possibilities of community energy generation projects.

All banks will go through a process called "due diligence" where they – and their lawyers – review all the documentation and contracts to make sure the risks and liabilities are understood and appropriately arranged. This covers rights to use the land on which the project is sited (and rights to get to that land), construction contracts, equipment-supply contracts and warranties, the power purchase agreement (with whoever has agreed to buy the energy output), the structure and governance of the company, operation and maintenance arrangements, planning permission and any associated conditions. And the bank will pay an engineering consultant to assess whether the design and expected output of the project is sound. This process can take a long time, be very

Part 3

expensive and will ultimately be something that you will pay for.

Whoever provides you a loan for your project will put considerable conditions on the loan. This will usually include putting in place legal rights to "step in" to take over the project if the company fails to repay the loan. The company's shareholders will lose their equity investment, but the bank takes the risk that, in such circumstances, it will be able to recover its loan from the continued operation or sale or the project.

As mentioned in Chapter 4, another source of loans, particularly for the risk money needed at the beginning of projects, is government money in the form of UCEF, RCEF and, in Scotland, CARES. These loans have the advantage that they don't need to be repaid if a project fails. In Scotland there is also government loan money available called REIF to fill capital funding gaps that commercial lenders are unwilling or unable to fill.

# PART 4
# COLD, HARD REALITY AND WHAT MIGHT BE

The hard truth is that most community energy projects still fail. This final part of *Community Energy* firstly looks at some of the reasons why so many projects don't succeed. It examines obstacles for projects, including practical issues that cause problems, like planning permission, lack of money and lack of land. It also looks at some of the problems that can arise within your group, how the complexity of projects can derail them and the role that politics can play. These problems are then illustrated by a couple of projects that have struggled to succeed despite all the goodwill and efforts put into them by those involved. By way of contrast, the final chapter then imagines what the future might look like in one community.

Chapter 16

# WHY IS IT SO DIFFICULT?

This book has taken a deliberately positive view of community energy – projects bring many benefits to local communities and the people who live in them, and community energy projects are realistic and achievable – but it would be wrong to give the impression that *all* community energy projects happen or that getting a project off the ground is easy.

Many projects never make it beyond the stage of being an interesting idea. Apart from the simplest, those that do make it further can often be fraught with problems, take a long time and, sadly, fail at one of the many hurdles they have to cross and never reach fruition. There are no records that tell us what percentage of projects fail* but the anecdotal evidence is that it's certainly more than 50%.

This chapter is therefore the slightly depressing one that looks at some of the many reasons why projects fail and also considers some examples of projects that have failed to progress despite having significant amounts of both goodwill and effort put into them.

---

\* If nothing else, it's quite difficult to define when a project has actually started as distinct from being an interesting idea that's just a subject for discussion..

## Practical issues

There are any number of practical issues that can derail a project. Here are some of them:

### Planning permission

The majority of energy generation projects will require planning permission from your local authority. Failing to get planning permission is a problem that many projects have and obviously one that will completely halt a project.

It might well be that there are no obvious issues with the location you're looking at for your wind turbine or hydro scheme, but that doesn't mean that it will fit with the planning policies your planning application will be judged against. One of the things you need to do early on in your project is understand those planning policies and how well your project fits with them.

From media coverage over the past few years, it would be easy to conclude that there are few planning constraints on where wind turbines, for example, can be built. In fact the opposite is true: local authorities have generally developed policies that mean that the areas where wind turbines can be built (often referred to as "areas of search") are highly constrained.

It's also important to understand that just because your project will be good for the community (as well as being good for the environment), it won't be judged any differently or, at best, only slightly differently than an out-and-out commercial project. Working with your local authority is discussed in Chapter 15 and, regardless of the rights or wrongs, it's important that you don't approach the council officers with the view that your project is entitled to be approved (or, if that is your view, you don't let it show too much!).

Even if your project fits with the planning policies of the council, that doesn't mean that it will be granted planning permission. There are any number of other issues that need to be satisfied, and, if they're not, your project will be refused permission. These include:

- Local ecology – will your plans cause problems for any wildlife in the local area, particularly protected species or any rare plants?

- Visual impact – even if your site is in an "area of search", the planning officers might still conclude that the visual impact of what you're proposing is too great.

- Airport or air traffic control radar – wind turbines can cause problems for radar systems even if your site isn't near or on a flight path for an airport.

- Noise – if your project has any housing nearby, you'll need to be able to demonstrate that any noise produced is below certain limits.

- Emissions – if your project is a wood-fuelled system, the emissions will need to be below prescribed values.

It's obviously worth hiring experts with the expertise and experience to address these issues and others like them, and help you through the planning process. If you're in the very fortunate position of having someone within your group who can do it for you, you should be aware of the amount of time and effort it will take them to do it.

Planning permission for many energy generation projects will be decided by the planning officers themselves, using what are called "delegated powers", but decisions on some projects, particularly if they're controversial or over a certain size, will be decided by councillors who sit on the council planning committee or equivalent. In these cases, the planning officers will make a recommendation to the committee, but ultimately it would be the committee that made the decision whether to grant planning permission or not.

Very often planning committees will follow the council officers' recommendation, but you shouldn't expect this to be the case. Plenty of renewable energy projects have been recommended for approval, only for the planning committee to refuse permission. It's therefore worth keeping your local councillors in touch with what you want to do. There are strict rules about what councillors who sit on the planning committee and will be making the decision can say or do, but there's nothing to stop you keeping them informed and nothing to stop you discussing the project with their colleagues either.

## Grid connection

If your project is generating electricity, then it will almost definitely need to be connected to the national grid. To do this you need to contact the Distribution Network Operator (DNO) that operates the local distribution grid for your area.

There can be two problems with grid connections that can derail projects. The first is a technical one where the local grid near to your project might not have enough spare capacity to take the electricity your project will be generating, or even that the distribution network close to your project is only single-phase and it would be impossible to connect your wind turbine or hydro scheme to the grid. If either of these is the case, then you could ask the DNO to upgrade the network in your area, but you will have to pay for this, which leads neatly on to the second grid connection problem.

Connecting your project to the local distribution network can be surprisingly expensive and you need to factor this into the overall capital cost for your project. If you fail to do this, you might be in for a nasty surprise just when you thought everything else had been sorted. Similarly, if you think the distribution network won't be up to dealing with your project and will need to be upgraded expect the work required to do this to be eye-wateringly expensive. DNOs are required to quote you a fair price for any work required, but there's usually some room for manoeuvre.

## Lack of renewable resource

It might seem obvious that you live in a windy area or that the river always has plenty of water flowing in it, but it's a mistake to make too many assumptions about these things. Plenty of projects have failed because, in fact, the location in which you want to put the wind turbine isn't as windy as you thought or it's the wrong kind of wind. Similarly, water levels in rivers and streams can vary dramatically depending on recent weather. If you're in an uplands area in particular, it might be that the river is a fast-flowing torrent soon after rain, but the rest of the time is much less dramatic.

For biomass projects, you need to think about where you are going to source the woodchip or wood pellets from. Are there suppliers nearby and where do they get their material from? Some woodchip is now being imported from other countries. If you want to manage the whole supply chain, then be aware of just how much woodland you need to supply woodchip on a sustainable basis.

## Lack of money

Nearly all renewable energy generation projects are capital intensive – that is they need a lot of money at the start of the project to install the wind turbine, build the hydro scheme, install the biomass boiler, and so on. This money has to come from somewhere. You might be in the amazingly fortunate position of being part of an organisation with reserves that you want to invest in the project, or you might be lucky enough to be able to source some as grants or donations, which don't need to be paid back. Otherwise the money will be being lent to the project from individuals or, for example, from banks.

Wherever you get the money from, you clearly need to make sure you will have enough of it. This might seem obvious, but be aware of the need for contingency funds. Unpredictable obstacles and problems are to be expected and some of them will need money to be overcome.

If you're borrowing the money, then you need to be sure that you'll be able to pay it back in a way that satisfies whoever is lending it to you in the first place. This might seem obvious but, again, you need to allow for things going wrong. For example, many microhydro schemes have had technical issues, which mean they haven't been able to operate for a period of months. This means you could have an extended period where you aren't generating any electricity and therefore you aren't generating any income either. If this happens, how do you plan to pay back your creditors?

For a project of any size, it's usually worth getting some financial modelling done by an expert. This should be able to show you how viable the project is financially, what the limits are on its viability, and what the chances are of these limits being tested (e.g. if it doesn't rain for six months, how will that affect your income and what are the chances of it not raining for six months?).

As well as needing capital to be built, many energy generation projects also need "seed money" or "at risk" money. This is necessary to get a project to the stage where it can be built and will pay for the planning application, legal agreements, financial modelling and so on. The problem at this stage is that this money is entirely "at risk" so if the project doesn't go ahead the money's been spent and there's no way of getting it back. Traditionally, this was a major problem for community groups who rarely had access to this kind of money. However, as described in Chapter 4, government schemes such as CARES, RCEF and UCEF are now in place so, for the moment at least, there are solutions to this problem.

## Failing to establish an agreement on the land

It's important early on to establish who owns the land you want to use for your project. Plenty of otherwise viable projects have failed to progress because it hasn't been possible to achieve a land agreement – clearly, if the landowner is not interested, then the project can't go ahead in that location.

Land ownership in the UK is sometimes not a straightforward thing to establish, with land sometimes being owned by offshore companies and the like. It can be surprisingly difficult sometimes to establish who owns a piece of land or who you should be contacting to have an initial conversation with. If you can't establish who the owner is, you can't go ahead in that location.

Reaching agreement on the land for a project can take a long time, and discussions with the landowner will often have to take place in parallel with other activities. Particularly in rural locations, it's highly likely that you won't be dealing directly with the owner of the land at all but with their land agent, which can add an extra complexity into discussions and negotiations. Try and involve the landowner directly in the discussions if you can, although sometimes that simply won't be possible.

Part 4

## Your group

Having looked at most of the practical problems that can derail projects, let's look at other, less tangible issues that can cause problems.

### Inexperience

If this is the first project your group has attempted, the chances are you won't have much relevant experience or knowledge within the group. For many of the technical, legal and financial issues that you are likely to come across, this lack of experience can be compensated for by hiring consultants who do have experience in these areas (provided you can find the money to pay for them, of course). The most likely outcome from a lack of experience is that you propose a project that is simply impractical. You'll probably discover the problems with the project as soon as you talk to a professional, and hopefully you'll be able to change direction before it has gone very far.

What can be more of a problem with inexperience is a lack of understanding of the overall process, how long it's likely to take and what you're likely to get out it. This can result in the group having false expectations, which later leads to disappointment and disillusionment. Even worse is when false expectations are communicated outside the group, to the wider community. You shouldn't underestimate the time projects can take and how difficult it will seem at times. Almost every project, whether ultimately successful or not, will have "war stories" about the bad times when it seemed like the project had hit the buffers.

The other side to inexperience is that you might find yourself asking the naive but penetrating questions that those with more experience wouldn't think to ask. These "daft" questions can sometimes be very useful.

### Time and effort

As mentioned above, community energy projects can take a considerable amount of time to come to fruition – a number of years usually – and a great deal of effort as well. This time and effort can be wearing on the individuals within the group, who will remember at some point during the process that they used to have social lives they quite enjoyed. Like most things to do with group dynamics, there's no simple answer to this except to make sure everyone understands from the outset that it might take some time for the project to be completed, that everyone has other pressures on their time and that you're all prepared for people dropping out because of the other commitments they have.

## Project complexity

Renewable energy generation projects are complex and, apart from the smallest, will inevitably involve multiple stakeholders who all need to be managed if the project is going to succeed. At the absolute minimum, there is likely to be:

- Your community group and its members.
- The landowner for the site of the project.
- The people or organisations funding your project.

Relationships need to be formed between each of these, and ultimately legal agreements defining the rights and responsibilities for each within the context of the project need to be drawn up.

On top of those, there will be the local authority that has already been discussed and your local DNO providing the grid connection. You are then likely to have a number of consultants working on your project over its lifetime – each of these needs to be managed – and, of course, a firm of lawyers who will be drawing up the various legal agreements that will be required.

You are also likely to be working with relatively large sums of money – hundreds of thousands if not millions – and signing agreements defining what these large sums of money will be used for and how they will be paid back.

This complexity and the responsibilities involved are other factors that can feel too much for a group of volunteers to deal with. What seemed like a straightforward renewable energy project at the outset can feel bogged down in a morass of stakeholders each needing to be satisfied, decisions that feel like they have nothing to do with the original goals, and responsibilities that are too big to be taken on in your spare time.

# Politics

Most community energy projects of any scale will rub up against politics at some stage, particularly if they're going to have a significant effect on local communities.

## Local politics

Renewable energy generation projects can often be controversial, particularly if they involve wind turbines. If your project does involve a wind turbine, there will definitely be people in your community who will have concerns about it. Many of these concerns will be legitimate, but it's also true that anti-wind-farm campaigners are extremely well organised these days and will often look to amplify any concerns and mobilise support for their cause within your community.

This can completely derail your project and, at its worst, can result in a community being split, with considerable ill feeling on both sides. Not surprisingly, many people who might want to be supportive will be afraid to do so publicly in this kind of environment. Your group may feel that it wants to continue and battle through this, but it's worth considering what the long-term cost might be and whether a change of tack might be a better approach.

## Government policies

For a period, in a relatively small way, central government had been increasingly supportive of community energy (see Chapter 4), although the same wasn't true for renewable energy more generally, with support being withdrawn or reduced, often with little or no warning. However, since the elections in May 2015, the UK government has shown little interest in community energy and, as has been mentioned a number of times, renewable energy is a significant part of most community energy projects, so government policy towards

Part 4

renewable energy has a huge effect on community energy groups.

The continual changes in government policy can be a particular problem for community energy projects that take a number of years to get up and running, because it means that at the outset you can't be sure if your project will stack up financially by the time it's ready to go. Some of this uncertainty is even built in to the FIT and RHI policies in the form of degression. This mechanism is used to reduce the value of FITs and RHI payments based on how many installations there will have been in the previous period. The problem with this is that it's impossible to predict accurately what the values will be from, say, two years before a project goes live.

While degression is a problem for projects, it can be managed to some extent. What's more difficult is when support for a technology or a particular scale of technology is withdrawn completely and suddenly. This is obviously a problem for specific projects, but also creates uncertainty more generally, so groups might decide not to continue with a project because they don't have enough faith that the necessary support to make it viable will still be there in the future.

Since 2010 there has been a great deal of uncertainty surrounding government policy towards renewable energy and energy-saving initiatives. This uncertainty is in itself damaging. Like any organisation, community groups need to manage the risks inherent in projects. The lack of clarity and certainty in government policy towards renewables reduces the chances that community groups will be prepared to take on projects that typically take a number of years to come to fruition.

## Government focus

Most government focus on renewables in the UK has been on business and, to a lesser extent, individuals in the home. Although there are exceptions, particularly in Scotland, the UK government's policies on community energy and community activity have been an afterthought and a case of trying to fit some community energy policy into existing frameworks. This is particularly true of the Renewables Obligation Certificate (ROC) and Contract For Difference (CFD) mechanisms, which are squarely aimed at encouraging business to invest in renewable generation. This is a result of the fact that consecutive governments have taken the view that investment from business is the way to achieve growth in the renewables sector. In a sense it's a bit like the Private Finance Initiative (PFI) approach to building hospitals and other public buildings, where government avoids the capital expenditure, but pays for the private investment over the long term.

This has several consequences. It means that community groups can find themselves trying to use mechanisms that were really intended for business and are an uncomfortable fit for the different ethos and approach of community organisations.

## Specific to community energy

As the two case studies in this chapter demonstrate, one of the problems for community energy groups can be working with the larger organisations that they naturally come up against while developing projects. There are a number of reasons why these relationships can be problematic. The first is that community groups can find that they're not taken seriously. They're usually very small and operate on a mostly voluntary basis. The projects they'll be looking to develop are themselves very often small (while an 80kW hydro scheme costing £750,000 is large by community standards, in the industrial world, it's tiny). It's easy for large organisations to view them as being "amateurs" in the worst sense of this word and to be viewed as irritants who are more trouble than they're worth.

This is compounded by the second problem, which is one of cultural differences. Community groups are very often enthusiastic, innovative, entrepreneurial and willing to take risks. Large organisations, whether commercial or governmental, will often be completely the opposite, with much of the focus both for the organisation and for the individuals within them being on procedures, processes and avoiding risks. These cultural differences can lead to significant misunderstandings on both sides and ultimately a breakdown of trust.

The ideal for the community group in these situations is if they can find a "champion" within the larger organisation who gets what they're trying to do and is willing and in a position to help them make it happen. This is, of course, often easier said than done.

# Case Study 1: Sheffield Renewables – Jordan Dam

Sheffield Renewables was formed by a group of volunteers in late 2007. The main aim of the group is to improve the city's environmental sustainability. Like many other community energy groups, they are concerned about the impacts of climate change and fossil fuel depletion. The group was formed as a social enterprise so it operates as a business with any surplus earnings from projects being re-invested to support new work as well as benefiting the people and communities of Sheffield.

## Jordan Dam

One of the first projects the group proposed was a hydro scheme at Jordan Dam on the river Don next to the site of Blackburn Meadows Wastewater Treatment Works. The group had been inspired by the community hydro project at New Mills in Derbyshire and believed that something similar should be possible in an urban setting. A desktop study indicated that there were five or six possible sites for a hydro scheme in Sheffield and the best of those was at Jordan Dam. A feasibility report produced in 2010 indicated that a hydro scheme at this location would indeed be feasible.

The scheme that was proposed would have used an 80kW Archimedes screw turbine producing around 310MWh of electricity a year. As discussed in Chapter 7 on hydro schemes, Archimedes screw turbines work well where there isn't a large fall, but there is the constant flow of a large volume of water.

There were three main stakeholders that the Sheffield Renewables group had to work with on the site: 1) Yorkshire Water who owned the site itself, 2) the Canal & River Trust who operate the canal, which starts at the dam, and 3) the Environment Agency. One of the challenges of the project was to manage the relationships with each of these organisations who all had their own interest in the site and who were all significantly bigger organisations than Sheffield Renewables.

Agreement was reached with Yorkshire Water who owned the site. They were happy for the project to be at the site and they also indicated that they would buy the electricity produced by the scheme to use in the waste-water treatment works.

A planning application was submitted to the local authority in February 2011 and planning permission was granted in May 2011. In 2012 the group developed a business plan for the scheme based around raising a significant part of the capital required from a local share issue. In the same year, the group held pre-tender interviews with six contractors interested in developing the scheme and the project was put out to tender late in 2012. Tenders were received from two contractors early in 2013.

## Problems

However, two significant issues had come to light during the latter stages of this process that were causing problems for the project. Firstly, an outfall sewer from the wastewater treatment works was found to run underneath the site which hadn't been spotted as part of the feasibility study. To obtain sitemaps showing exactly where it was and what it was from Yorkshire Water and Sheffield Council took over six months. The existence of the sewer would mean that additional civil engineering work would be necessary.

The second problem was uncertainty regarding the design of fish passes required by the Environment

Agency. It only became apparent very late in the day that the already submitted fish pass design might not satisfy the Environment Agency who suggested that a second one would be necessary so that the scheme would comply with the new Water Framework Directive.

## A difficult decision

Both of these problems led to uncertainty over the tender and likely large price increases. The original estimate for the cost was between £650,000 and £750,000, but the lowest tender submitted was £850,000. This could only be reduced if Sheffield Renewables took on significantly increased risk.

The group were faced with a very difficult decision at this point: whether to proceed with the project with the additional risk or whether to suspend the project despite all the effort and commitment that had gone into getting it this far.

Ultimately, the group decided that the additional risk was too much and agreed to mothball the project. As it says on the Sheffield Renewables website:

> The decision to recommend suspending work on the project has been a difficult one, particularly after all the hard work which has gone in over the last five years. If Jordan Dam Hydro does become viable in the future, we would still hope to go ahead.[1]

Ultimately this project had proven too much for a community group to manage. The technical challenges of a hydro project were compounded by the challenges of dealing with the different stakeholders on the site and all of this done by volunteers, in their own time. As well as being time-consuming, the project had also been emotionally draining.

## A new approach

However, this particular story does have a happy ending. After the decision had been taken to suspend the work on the hydro project, the group decided to refocus their activities on solar pv projects. The aim of this new initiative was to install panels on the roofs of buildings with strong community links. Their view was that this type of solar pv project is significantly less complex, will involve fewer risks and many of the original aims of the group could still be achieved.

By the time the decision had been taken to suspend work on the hydro project, the group had already raised around £240,000 from people keen to invest in the project. Because the project hadn't gone ahead, none of this money had been spent. The group offered to return it to the investors, but were very pleasantly surprised when the vast majority of them said that they would rather Sheffield Renewables kept the money and invested it in the solar schemes instead.

So far three 50kW solar PV installations have taken place. The first on Paces Campus was commissioned in April 2014; the second at Swinton Fitzwilliam primary school commissioned in October 2014; and the third installation at Attercliffe police station in 2015. Towards the end of 2016 they had a share offer that raised £70,000 to finance further installations and are currently looking for suitable sites for these.

Everyone wins out of these projects with Sheffield Renewables able to sell the electricity generated to the host-buildings at less than the market level; the investors receiving a modest return on their investment; and Sheffield Renewables able to make a small profit that's reinvested in other schemes.

# Case Study 2

This Community Energy Project has many features in common with the Jordan Dam project in Sheffield with the community groups showing remarkable tenacity and resilience, but ultimately there being too many obstacles to make the project work.

The project was initiated by a community organisation formed in 2005 in a city suburb as an offshoot of a group that had successfully campaigned against a large supermarket being built in the area. A number of people in that campaign had a wider interest in localism and the green agenda. After the supermarket campaign ended an organisation was set up to pursue these interests.

Inspired by rural communities creating energy initiatives using wind turbines, the group were interested in the possibility of building and owning a wind turbine even though they were in an urban area.

## The first site

The first site the group looked at for a turbine was just next to the coast behind a bus depot. A firm of consultants were commissioned to produce a study funded by a government initiative. This study indicated that the site

Part 4

was feasible. Additionally the local council were willing to lease the site for the turbine.

However, some problems with this scoping study were found. Another firm with more experience were asked to review the work and found: a) a turbine on the site would have interrupted a key view from the city's historic centre; b) the site was next to a Special Protection Area for bird species and; c) the site was subject to coastal flooding. Combined, these factors meant that there was little chance of getting planning permission for a turbine on the site.

## The second site

However, the new consultants also identified a possible alternative site up the coast slightly, on a spit of land that is part of a waste water treatment works that was operated under a Private Finance Initiative (PFI).

Although the owners of this new site were initially enthusiastic about the project, stating they would purchase the electricity produced, everything seemed to slow down when it came to drafting the legal agreement on the land the project would occupy. Eventually, it transpired that there was another stakeholder that the groups hadn't previously been aware of: a company that had invested significant sums in the sewage works under the PFI agreement and the agreement of this company was necessary for the project to go ahead. Unfortunately, this company had nothing to gain from the project which in their view would increase the risk of something going wrong on the site, with the potential to reduce their profits.

After 18 months of negotiation the community group had to conclude that this second site wasn't viable either.

## Thinking laterally

In the preceding months the group had won a competition. Through this they had gained significant public support for the project and a significant grant. As a result, they had enough money to look at other potential sites. However, their consultants advised them that there were no other possible sites for a wind turbine of the size they wanted to develop anywhere in the local area.

This is where the community group took a lateral step towards achieving their goals. They were approached by another firm of consultants, who suggested that they could look to build and own one in another part of the country.

The consultants suggested a number of sites for a relatively small commercial wind development. After reviewing the risks and opportunities, the community group chose a suitable site that had a sympathetic landowner. A planning application was submitted in 2014. Sadly, planning permission was refused. The community group appealed this decision but their appeal was also unsuccessful.

## The end of the line

The group felt they had now reached the end of the line and couldn't put more time and effort into the project, particularly when government support for renewables was being reduced and would be reduced further. Reluctantly, they took the decision to end the project.

Chapter 17

# THE FUTURE

Attempts to predict the future are always more likely to fail than to succeed. This chapter is therefore not going to make predictions as such, but is going to look at a couple of directions community energy could go in.

At the moment, it feels like we're at a cusp: community energy is now recognisable as a movement in its own right. Over the years, the pace of growth of this movement has varied in the different parts of the UK. The greatest growth has tended to be in areas away from the centre, especially Scotland, where the devolved government under different administrations has been better at recognising the potential and putting in place policies that have enabled this potential to be realised.

To some extent, the rest of the UK has been catching up over the past few years, and the growth in the number of community energy generation projects in England has been good to see, together with the emergence of Community Energy England and Community Energy Wales (Ynni Cymunedol Cymru) and the Powering Up events in various parts of England. The nature of projects tends to be slightly different in different parts of the UK, with more cooperative-investment-based projects south of the border, but the aims and objectives – addressing issues like energy use, fuel poverty, sustainability, and climate change – remain broadly the same regardless of where you are.

## Government policy

As is clear from the rest of this book, the growth and the shape of community energy is highly dependent on government policies. This shouldn't be a surprise: for all that it's privatised, the energy industry in the UK, of which community energy is a very small part, is highly regulated and its direction largely defined by government policy. While the movement has grown and has some momentum of its own, it's still very small and is largely dependent on the time and goodwill of volunteers. It would be easy for it to fail to progress much further.

While there are specific government policies that have been aimed at helping community energy, it should also be clear that the community energy movement is affected by policies on renewable energy generation, energy reduction, fuel poverty and the need to address climate change. Any change governments make in these areas of policy have knock-on effects for the community energy movement.

Part 4

This is why recent announcements made by the Westminster government on renewable energy generation and energy saving have been troubling. These include:

- Ending support for commercial-scale onshore wind farms through both the Renewables Obligation scheme and the Contract For Difference scheme.

- The significant reduction in FIT rates and RHI rates.

- Ending support for commercial scale PV farms.

- The ending of the Green Deal scheme that provided financial support for householders wanting to make energy-reduction improvements to their houses.

While only few community groups will be affected by the announcements on commercial-scale wind farms, and few community energy groups had managed to successfully engage with the Green Deal scheme (which was riddled with so many flaws that it never achieved what it set out to anyway), the announcements on FIT and RHI rates have directly affected community projects. Furthermore, the mood in the renewables industry created by these announcements, combined with other announcements on unconventional gas exploration (fracking), is not good and the implication is that this is a government that's not keen on renewables and therefore not particularly in tune with the aspirations of community energy.

However, it's worth remembering that community energy shouldn't just be about renewable energy generation projects and bringing money into a community. Communities can tackle climate change and address local fuel poverty in other ways, and while energy generation can continue to be important, community energy groups can also look at other ways to achieve their aims.

## Energy reduction

One area that a significant number of community groups have been involved with in the past is energy reduction, but it seems like there's much more in this area that community groups have the potential to achieve. As noted above, one of the recent announcements from what was the Department of Energy and Climate Change (DECC) was the closure of the Green Deal scheme. The scheme had never achieved what was intended, with only 15,000 Green Deal[1] loans being approved over its lifetime, but currently there is no announcement of any replacement.

Although the energy requirements for new housing are good, the UK housing stock is poor and there is the well-documented problem of households in fuel poverty, some having to choose between heating and eating – it's estimated that six million low income homes have an energy rating of D or worse.[2] These and other facts like them make it likely that a replacement scheme for the Green Deal will be created at some point.

Community energy groups could play a significant role in any new energy reduction scheme, provided it is designed so that they can successfully engage with it and be agents for it. As the various case studies in this book hopefully demonstrate, community energy groups have a desire and the motivation to want to see significant improvements in the energy efficiency of the houses in their area. They also have an entrepreneurialism that would enable them to make the best use of any opport‐unity they're given in this area. The Reepham Insulation Project described in Chapter 1 is a good example of how a community group can deliver significant improve‐ments to household insulation and energy efficiency.

One of the key ingredients for allowing community groups to help deliver a new scheme would be providing enough funding to support the employment of commu‐nity-based energy advisors. Volunteer time and effort can only go so far and while it's right that community energy groups are primarily led and governed by volunteers, when it comes to delivering projects on the ground, like significant community energy reduction, there is a need to bring in people who can devote their time to the work.

These energy advisors wouldn't do the work of fitting the measures, but could provide whatever help is required to nudge householders considering energy improvements so that they go actually go ahead and don't get halted by what are often surmountable hurdles. These advisors could, for example:

- Advise on what steps are appropriate for each home, taking into account the lifestyle and needs of the homeowner. Even if the structure and build of the house is similar, the needs of a retired couple are going to be very different from the needs of a young family.

- Help householders negotiate their way through the rules of the scheme and help them understand what the various options are. Whatever shape the new scheme takes, there will doubtless be an application process and various steps to take. Even simple processes can be enough to stop someone going ahead. Having an advisor on hand who is experi‐enced in the process and can talk them through it could be enough to keep someone on track who might otherwise decide it's not worth the effort.

- Help householders choose between different local suppliers, depending on what their needs are. Many people would be happy doing this themselves, but again, for some people having somebody who is unbiased and experienced to hand could make all the difference in whether or not they go ahead with improvements.

There are clearly other aspects of whatever shape the new scheme takes that would help community energy groups engage with it, but having a facility for local energy advisors could be a significant component in whether an energy reduction scheme is successful or not.

# Delivery and supply of electricity

The other area where community energy groups could make a difference is in the delivery and supply of electricity. Except for off-grid projects, up to now it has always been more or less impossible for community groups to get into this area. However, this is not the case in other parts of Europe and particularly Germany.

With an Energiewende* policy supported by all major political parties, Germany is often held up as the European poster child of renewable energy generation and more specifically locally owned energy generation.

Direct comparisons between Germany and the UK are difficult because there are so many other legal, financial and cultural differences between the two countries, but by the end of 2013 Germany had around 75 gigawatts (GW) of installed renewable capacity (around 40% of the total electricity generating capacity). Over 20% of this was owned by co-ops, community share groups and similar, while a further 25% was owned by individuals. By contrast, the UK had around 19GW of installed renewable capacity (24% of the total electricity gener–ating capacity) and around 0.3% was community owned. (The figures in the UK has been rising since then, but then so have the German figures.)

However, for all that the adoption of renewable electri–city generation in Germany is impressive there's more to what has happened, and some of it can give an indica-tion of what might be possible in the UK with the right government support and the right regulatory changes – a few of which are starting to happen.

* Energiewende (Energy transition) is the long-term policy in Germany to move away from fossil fuels and nuclear power to renewable energy, energy efficiency and sustainable develop-ment. It is motivated both by a need to address climate change and a distrust for nuclear power following the Chernobyl disaster and, more recently, Fukushima.

## Supply of electricity

Looking at Germany from the UK, what's as interesting as the high level of community-owned generation is that increasingly local ownership of the distribution grid and of electricity supply is starting to take place.

Over the years, both the UK and Germany split the electricity industry into various distinct elements, including generation, the transmission grid, the local distribution grid and the actual retail supply of electricity to customers. In the UK, so far, it has been more or less impossible for anything as small as a community group to gain ownership of the local distribution grid for the supply of electricity. This is one reason why the majority of community renewable projects have been based on generation – it's been the only part of the system that was available for groups to engage with.

The history of the German electricity industry is very different and this has enabled local co-operatives and similar groups to own and run the distribution grid, with 190 communities successfully doing so by the end of 2012. A small number are now also taking on the retail side as well, giving them complete control of local electricity supply. Having this control means that communities can generate and supply electricity locally so that the economics remain local, providing further incentives for developing local renewable generation.

A number of actual and proposed changes in the regulation of the UK electricity industry have started to make this kind of approach viable here as well. These include "Licence-lite", allowing organisations to piggy-back on an existing licensed supplier and enter the supply market, and "Non-traditional business models".

On the back of these and other moves, a number of community-led projects have been set up around the country to explore how community groups might be able to take advantage.

## Fintry and smart meters

One of these projects is in the village of Fintry in central Scotland. As well as having a well-established community energy group, Fintry Development Trust (FDT - see Chapter 1), several years ago a commercial 1MW anaerobic digestion (AD) plant was set up a few miles outside the village. This plant takes silage from local farms and waste from whisky distilleries as fuel and not only does it produce electricity but also fertiliser and animal bedding as by-products. The plant feeds the electricity produced into the local grid, but the owners, Strathendrick Biogas were always interested in the idea of delivering electricity locally.

FDT had also had a long-standing interest in supplying electricity locally, having first approached Ofgem to discuss the idea in 2008. At the time, they had been met with looks of blank incomprehension and had concluded that the notion was too difficult. Discussions between FDT and Strathendrick Biogas started soon after the AD plant had been built revealed this common ambition.

### Feasibility project

In the first instance a feasibility project has been completed to look at how the AD plant could deliver electricity to local consumers. Initially two models were looked at: the first would have involved setting-up a "private wire" to delivery electricity from the plant to householders in parallel to the existing network, but this was always going to be expensive and cumbersome. The other approach was to use the existing network and fit each house with smart meters and a control system on the AD plant, which could be used to monitor and manage consumption using active network management software.

As well as providing a much more detailed record of electricity consumption (down to minute by minute), some smart meters can also control domestic appliances, including heating systems, washing machines and tumble dryers. Smart meters of this type, combined with sophisticated monitoring and control on the generating equipment, means that you can have a "smart grid" where production and consumption can be matched.

In the case of Fintry, this "smart grid" aspect of the project is important because the owners of the AD plant want to increase production, but the local distribution grid has reached capacity. So in order to achieve the increase in production using traditional grid management techniques, the local grid would need to be upgraded, increasing the cost of the project to the point where it would no longer be viable. By creating a "smart grid" and matching consumption to production, this upgrade is no longer necessary because any additional electricity produced will be used locally, not getting as far as the connection with the wider grid.

Smart grids or Active Network Management (ANM) like this have already been used on a larger scale to increase the amount of renewable energy generation possible on the Orkney and Shetland islands, where the grid connection to the mainland and even between islands had reached capacity. There are other similar projects

Part 4

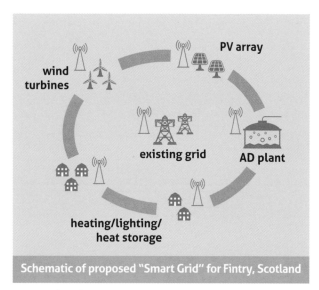

**Schematic of proposed "Smart Grid" for Fintry, Scotland**

testing the idea at a community level in the Scottish Borders and also on the islands of Mull and Iona.

### The plan

The proposal in Fintry is therefore to fit households with smart meters and also, where necessary, switch their heating systems to heat pumps. ANM software would be used to manage the electricity consumed from the local grid by the households and make sure that this consumption matches the production from the AD plant. The households would pay a "renewable energy" tariff for the electricity that they use to a company operated by FDT.

At the moment this is only a proposal and there are many hurdles to cross before it becomes a reality, but it does have some clear benefits for everybody involved:

- The cap on the amount of electricity Strathendrick Biogas can produce as a result of local grid limitations is removed, allowing a greater amount of renewable electricity to be produced.

- The operator of the local grid doesn't have to deal with increasing the capacity.

- Householders would gain reduced electricity bills as a result of having more efficient heating systems and having a controlled electricity system.

- It's estimated that the project could save almost 1,000 tonnes of $CO_2e$ per year.

- It retains more of the money spent on energy in the local area rather than it going to one of the big energy providers.

This type of project may be one direction that community energy groups start to go in – becoming local retail energy companies and as a result providing direct fuel-saving benefits to householders in their area. This could be particularly beneficial if community groups are also able to play a role in delivering significant energy-reduction measures.

# A DREAM REALISED

*If it was late on a Sunday, she would walk up the hill. It might be the best day of the year so far, sunny and almost warm, but it was now afternoon and the shadows would be lengthening and the air chill; the light warm on the rock. This last bit of the climb was always the most enjoyable: up across the almost-flat top, the path easy between the tussocks and, now that she was out of the lee of the hill itself, the breeze strong in her face.*

*However, she would also enjoy these moments for other reasons. She knew if she turned round, she would already be able to see the village itself down in the valley and the hills on the other side. She would wait though and not look until she was actually at the top, savouring the moment and the knowledge of what she would be seeing (a childish anticipation maybe, but you had to get your pleasures where you could).*

*Earlier in the afternoon, before leaving the house, she would have checked the production figures and noted them down in the spreadsheet like she always did. It would probably have been a good week; not the best, but about average for the time of year. Her partner would like to gently mock her for this habit, but she was unrepentant: it mattered to her even though she would admit that there was no practical value in it.*

*She'd be at the top of the hill now and would have a ritual for this moment: walking up and touching the frost-bitten and crumbling concrete trig point, before wandering over to the cairn a few yards away; still refusing to raise her eyes to the view; still enjoying the anticipation. Only after she sat down and caught her breath would she bother to look up and out. And like it always was, across the valley, the turbine would be there in front of her, the blades gently turning.*

*This was what had started it: the turbine on the hill. It had seemed fanciful at the time, a village out in the middle of nowhere owning and running its own wind turbine. What was the point of that? She tried to avoid using the term, but it was quixotic to say the least. There it was though, turning away as it had most days for the past five years, and everything else that had happened had happened as a consequence of that first project. But that was another phrase she tried to avoid. It sounded too much like a line of dominoes falling over after the first one is given a gentle push. Nothing could be further from the truth. The first project had taken much more than a gentle push, and while subsequent projects hadn't required quite so much effort, they had been far from easy.*

*Looking down at the village now, she could see the lights starting to come on – helping to map out the streets. After the turbine, the streetlights had been the next project. They had been lucky with that one. The council had been looking for an area to trial replacing the old sodium lights with the new LED lights and the village had been the right size and the street-lights had the right fittings for the new bulbs.*

*She sat down on one of the rocks and continued to look out over the village. She could identify each of the houses they had insulated, including her own. Initially it had been the volunteers who had gone house to house explaining what would be done before the job had become too big and they'd had to employ their first energy manager. Now even most of the hard-to-treat homes had been insulated and most of the houses had heat pumps.*

*The new deal on the turbine plus the smart meters would mean that for the first time they could use the electricity it produced locally. She admitted that she didn't really understand how the smart meters worked, but it meant more money staying locally rather than going to a big energy company, so that was good.*

*She stood up again and looked down the valley where she could see the lights of the neighbouring village and still just see through the dusk the ribbon of river that was maybe, possibly, hopefully the site of the next project.*

# REFERENCES

## Chapter 1

1   Isle of Gigha: GREL - The Dancing Ladies
    http://www.gigha.org.uk/windmills/

2   Ashton Hayes Going Carbon Neutral
    www.goingcarbonneutral.co.uk/background/

3   Hopkins R. (2008). *The Transition Handbook: From oil dependency to local resilience.* First edition. Green Books: Dartington.

4   Transition Linlithgow
    http://transitionlinlithgow.org.uk/

5   Transition Network
    https://www.transitionnetwork.org/

6   Department of Energy and Climate Change (DECC) (2014). "Community energy in the UK: Part 2. Final report",
    https://www.gov.uk/government/uploads/system/uploads/attachment_data/file/274571/Community_Energy_in_the_UK_part_2_.pdf

7   Scene
    http://scene.community/

## Chapter 2

1   Schiffer, A. (2014). "Community power Scotland: From remote island grids to urban solar co-operatives". Friends of the Earth Scotland,
    http://www.foe-scotland.org.uk/sites/www.foe-scotland.org.uk/files/CommunityPower%202.pdf

2   Schlömer S., T. Bruckner, L. Fulton, E. Hertwich, A. McKinnon, D. Perczyk, J. Roy, R. Schaeffer, R. Sims, P. Smith, and R. Wiser (2014) *Annex III: Technology-specific cost and performance parameters. In: Climate Change 2014: Mitigation of Climate Change. Contribution of Working Group III to the Fifth Assessment Report of the Intergovernmental Panel on Climate Change* Cambridge University Press, Cambridge, United Kingdom and New York, NY, USA., p.1335,
    http://www.ipcc.ch/pdf/assessment-report/ar5/wg3/ipcc_wg3_ar5_annex-iii.pdf

3   Department of Energy & Climate Change (DECC) (2013). "Greenhouse gas reporting: Conversion factors 2014", p. x
    https://www.gov.uk/government/publications/greenhouse-gas-reporting-conversion-factors-2014

4 Department of Energy & Climate Change (DECC) (2014). "Community energy strategy: Full report", https://www.gov.uk/government/uploads/system/uploads/attachment_data/file/275163/20140126Community_Energy_Strategy.pdf

5 Scottish Government (2009). "Scottish Index of Multiple Deprivation 2009: General Report", http://www.gov.scot/Publications/2009/10/28104046/0

6 Neilston Development Trust, "Renaissance Town Charter", http://www.neilstontrust.co.uk/what-we-do/town-charter/renaissance-town.html

7 Trading Economics, Brent Crude Oil, http://www.tradingeconomics.com/commodity/brent-crude-oil

8 Department of Energy & Climate Change (DECC) (2015). "Digest of United Kingdom energy statistics", https://www.gov.uk/government/uploads/system/uploads/attachment_data/file/450302/DUKES_2015.pdf

9 Capener, P (2014). "What is community energy and why does it matter?". Paper for Community Energy England, http://communityenergyengland.org/publications/

## Chapter 3

1 Schumacher, E.F. (1973). *Small Is Beautiful: A study of economics as if people mattered*. Blond and Briggs: London.

2 Centre for Sustainable Energy (CSE) (2009). "Switched on since 1979", https://www.cse.org.uk/downloads/reports-and-publications/switched-on-since-1979.pdf

3 The New York Times, "Global Warming Has Begun, Expert Tells Senate", http://www.nytimes.com/1988/06/24/us/global-warming-has-begun-expert-tells-senate.html?pagewanted=all

4 Boykoff, M.T. and Timmons Roberts, J (2007). "Media coverage of climate change: Current trends, strengths, weaknesses." Occasional paper, UN Human Development Report Office, http://hdr.undp.org/sites/default/files/boykoff_maxwell_and_roberts_j._timmons.pdf

5 Stern, N. (2006). *The Economics of Climate Change: The Stern review*. Cambridge University Press: Cambridge.

6 Wood, E. (2002). *The Hydro Boys: Pioneers of renewable energy*. Luath Press Limited: Edinburgh.

7 Baywind http://www.baywind.coop/about-us-3/

8 Energy4All http://energy4all.co.uk/about-us/

9 Community Energy Scotland
http://www.communityenergyscotland.org.uk/
about-us.asp

10 OFGEM - Public Reports and Data: FIT
https://www.ofgem.gov.uk/environmental-
programmes/fit/contacts-guidance-and-resources/
public-reports-and-data-fit

11 OFGEM - Public reports and data: Domestic RHI
https://www.ofgem.gov.uk/environmental-
programmes/domestic-rhi/contacts-guidance-and-
resources/public-reports-and-data-domestic-rhi

12 OFGEM - Publications library: Non-Domestic
Renewable Heat Incentive (RHI)
https://www.ofgem.gov.uk/environmental-
programmes/non-domestic-rhi/contacts-guidance-
and-resources/
publications-library-non-domestic-renewable-heat-
incentive-rhi

13 Department of Energy and Climate Change
(DECC) (2014), "Community energy strategy: Full
report",
https://www.gov.uk/government/uploads/system/
uploads/attachment_data/
file/275163/20140126Community_Energy_
Strategy.pdf

## Chapter 4

1 10:10 (2016). "Community energy: The way
forward",
http://files.1010global.org/Community-energy-the-
way-forward.pdf

2 Department of Energy and Climate Change
(DECC) (2014), "Community energy strategy: Full
report",
https://www.gov.uk/government/uploads/system/
uploads/attachment_data/
file/275163/20140126Community_Energy_
Strategy.pdf

3 OFGEM - ECO public reports and data
https://www.ofgem.gov.uk/environmental-
programmes/eco/contacts-guidance-and-resources/
eco-public-reports-and-data

4 Department of Energy & Climate Change (DECC)
(2015). "Green Deal and Energy Company
Obligation (ECO): headline statistics",
https://www.gov.uk/government/statistics/green-
deal-and-energy-company-obligation-eco-headline-
statistics-november-2015

5 Scottish Government (2015). "Community energy
policy statement – September 2015",
http://www.gov.scot/Resource/0048/00485122.pdf

## Chapter 5

1 Bovey Tracey Swimming Pool
http://www.boveyswimmingpool.org.uk

2 Microgeneration Certification Scheme
http://www.microgenerationcertification.org

## Chapter 6

1 BWEA - UK Wind Speed Database
http://www.bwea.org/noabl/

2    Sustainable Hockerton
     http://sustainablehockerton.org

## Chapter 7

1    UK National River Flow Archive
     http://nrfa.ceh.ac.uk/

2    Environment Agency
     https://www.gov.uk/government/organisations/
     environment-agency

3    Scottish Environment Protection Agency (SEPA)
     http://www.sepa.org.uk/

4    Callander Community Hydro Project
     http://www.callandercdt.org.uk/index.php

5    Whalley Community Hydro
     http://www.whalleyhydro.co.uk

6    Transition Town Clitheroe
     https://transitionnetwork.org/

7    Talybont-on-Usk Energy
     https://talybontenergy.co.uk

## 8   Chapter 8

1    Heat Pump Association
     http://www.heatpumps.org.uk/

2    Green Energy Supply Certification Scheme
     www.nef.org.uk/service/search/result/green-energy-
     supply-certification-scheme

## Chapter 9

1    Defra - Exempt Appliances
     https://smokecontrol.defra.gov.uk/appliances.php

2    Microgeneration Certification Scheme
     www.microgenerationcertification.org

3    Biomass Energy Centre
     www.biomassenergycentre.org.uk

## Chapter 11

1    Climate Friendly Bradford on Avon
     www.climatefriendlybradfordonavon.co.uk

2    Welcome to Green Streets – British Gas
     www.britishgas.co.uk/smarter-living/save-energy/
     green-streets.html

3    National Statistics, "English housing survey 2010 to
     2011: headline report"
     https://www.gov.uk/government/statistics/english-
     housing-survey-headline-report-2010-to-2011

4    Reepham Low Carbon Communities Challenge
     www.reephamchallenge.org/about

5    Velocoty Cafe and Bicycle Workshop
     velocitylove.co.uk

6    Carbon Conversations
     www.carbonconversations.org

## Chapter 12

1 Ofgem - FIT Tariff Rates
https://www.ofgem.gov.uk/environmental-programmes/fit/fit-tariff-rates

2 Ofgem - Feed-in Tariffs deployment caps reports
https://www.ofgem.gov.uk/environmental-programmes/fit/contacts-guidance-and-resources/public-reports-and-data-fit/feed-tariffs-deployment-caps-reports

3 Ofgem - Domestic Renewable Heat Incentive
https://www.ofgem.gov.uk/environmental-programmes/domestic-rhi

4 Ofgem - Non-Domestic Renewable Heat Incentive
https://www.ofgem.gov.uk/environmental-programmes/non-domestic-rhi

5 DECC (2013). "Investing in renewable technologies: CfD contract terms and strike prices",
https://www.gov.uk/government/uploads/system/uploads/attachment_data/file/263937/Final_Document_-_Investing_in_renewable_technologies_-_CfD_contract_terms_and_strike_prices_UPDATED_6_DEC.pdf

## Chapter 13

1 Foundations of Co-operation Rochdale Principles and Methods
www.uwcc.wisc.edu/icic/orgs/ica/pubs/review/vol-88-2/5.html

## Chapter 14

1 MCS (2012). Guide to the Installation of Photovoltaic Systems,
http://bpva.org.uk/media/38266/new-guide-to-installlation-of-pv-systems-mcs_20130530161524.pdf

2 Ofgem - FIT tariff rates
https://www.ofgem.gov.uk/environmental-programmes/fit/fit-tariff-rates

## Chapter 16

1 Sheffield Renewables
https://www.sheffieldrenewables.org.uk/

## Chapter 17

1 Department of Energy & Climate Change (DECC) (2015). "Green Deal and Energy Company Obligation (ECO): headline statistics",
https://www.gov.uk/government/statistics/green-deal-and-energy-company-obligation-eco-headline-statistics-november-2015

2 Labour Party (2014). "An end to cold homes: One Nation Labour's plans for energy efficiency",
http://www.cibse.org/getmedia/ae411f05-1a1e-4510-be3c-192f7141c684/An-end-to-cold-homes-Consultation-document.pdf.aspx

# INDEX

green books

www.greenbooks.co.uk

**Environmental publishers for more than 25 years**

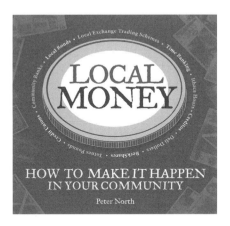

*"This book tells every community everywhere how to make local money work for local good"*

**Polly Toynbee, columnist,
*The Guardian***

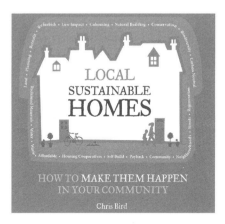

*"This book will inspire all concerned in making our buildings and communities fit for the future"*

**Penney Poyzer, broadcaster & campaigner**

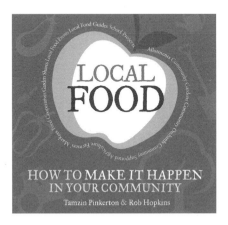

*"If you believe that real change happens from the ground up, start with food and read this book"*

**Patrick Holden CBE, Director,
Soil Association**

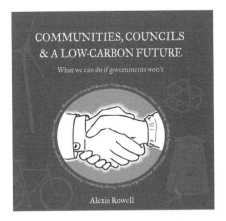

*"An invaluable resource for any green-minded councillor or community group"*

**Eugenie Harvey, Director,
10:10 Campaign**

green books

www.greenbooks.co.uk

**Environmental publishers for more than 25 years**

*"This has got to be the most practical and beautifully illustrated book on earth building ever published"*
**Keith Hall, editor, *Building for a Future* magazine**

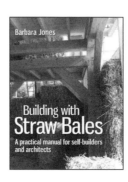

*"Offers architects and would-be builders a detailed understanding of how and why to build with straw bales"*
**Ben Law, eco-builder and author**

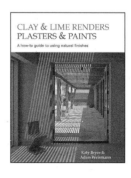

*"Adam and Katie are rethinking how we build and coming up with something beautiful"*
**John Vidal, environmental editor, *The Guardian***

*"A must-read for any student or home designer who seeks the wisdom necessary to work with nature's forces"*
**Sue Roaf, author of *Ecohouse***

*"A no-nonsense and engaging introduction"*
**Kevin McCloud, *Grand Designs***

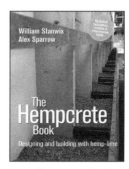

*"An excellent practical guide"*
**Jon Broome, author of *The Green Self-Build Book***